THE A1 STEAM LOCOMOTIVE TRUST

TORNADO

This book is dedicated to those people who, in 1990, had a vision to build and operate a class of main line express steam locomotive which had become extinct in the 1960s. Also, to all the individuals and organisations who have contributed in terms of time, expertise, and finance in turning that vision into reality.

Written in conjunction with the builder, owner and operator of No. 60163 *Tornado*

The A1 Steam Locomotive Trust
Darlington Locomotive Works
Hopetown Lane, Darlington DL3 6RQ
website: www.a1steam.com

Further information on how to become involved with *Tornado* is available by e-mail: enquiries@a1steam.com

First published in hardback in a larger format in 2011 as
The A1 Steam Locomotive Trust Tornado Owners' Workshop Manual
Reprinted 2012
Published in paperback 2015
This edition published in 2018

A catalogue record for this book is available from the British Library

ISBN 978 1 78521 573 5

Library of Congress control no. 2018948556

Published by Haynes Publishing,
Sparkford, Yeovil, Somerset BA22 7JJ, UK.
Tel: 01963 440635
Int. tel: +44 1963 440635
Website: www.haynes.com

Haynes North America Inc.,
859 Lawrence Drive, Newbury Park,
California 91320, USA.

Front cover photograph © A1 Steam Locomotive Trust/David Chandler;
back cover supplied by Alamy.

Printed in Malaysia.

THE A1 STEAM LOCOMOTIVE TRUST
TORNADO

Geoff Smith

HAYNES **ICONS**

Contents

Introduction

The year 1968 will be for ever etched into the minds of steam railway enthusiasts as the year when scheduled steam operations on British Railways came to an end. The dash to rid the network of steam led to the premature demise of many steam locomotives – indeed, the last main line steam locomotive to be built in the UK (Riddles Class 9F No. 92220 *Evening Star*) only emerged from Swindon Works in 1960.

OPPOSITE At Sheffield, *Tornado* reverses into a bay platform prior to the East Midlands Trains 'Meridian' unit also being named *Tornado*. *(Ian MacDonald/A1SLT)*

Certain locomotives were designated for preservation as part of our national heritage. However, these were far too few in number, but the speed at which British Railways sought to distance itself from the steam days left little time for enthusiasts to raise the money needed to take more into private ownership.

Some were successful, probably most notable among these being the Gresley A3 class 4-6-2 No. 60103 (or No. 4472 in its LNER days) *Flying Scotsman*. A large number of the locomotives sold for scrap to private breakers' yards went to Woodham Bros at Barry, South Wales. A combination of other work taking priority, plus Dai and Billy Woodham's own love of steam, resulted in delays in cutting up many of these locomotives. This gave the preservation movement time to mobilise and resulted in more than 200 locomotives being saved from the torch. Even today, some Barry 'wrecks' are still being restored, and the late Dai Woodham himself has become a hero within the steam locomotive movement.

Sadly, this approach did not extend to other scrapyards, and as the Peppercorn A1 class 4-6-2s mostly went to yards along the east side of the country, all were quickly reduced to scrap metal. The last Peppercorn A1 to be withdrawn was No. 60145 *Saint Mungo*. An attempt was made to take this into preservation, but in the end, A2 class No. 60532 *Blue Peter* was chosen instead.

The total elimination of one of the icons of the East Coast Main Line (ECML) steam era devastated many enthusiasts, a feeling which did not diminish as the years passed, and so began a story...

Acknowledgements

Over the course of the preparation of this book, the author has spoken to many individuals, and their input has contributed variously to its content. They are too numerous to mention individually, but grateful thanks go to all. I hope you all think the result was worth it!

However, there have been a number of key individuals, without whose invaluable assistance this book would not have been possible, so special thanks go to:

Mark Allatt, Gordon Best, Tony Broughton, Graeme Bunker, David Champion, Phil Champion, Malcolm Crawley, David Elliott, Ian Howitt, Tony Lord, Ian Matthews, Kevin Meghen, Rob Morland, Graham Nicholas, John Pridmore, Ian Storey, Alexa Stott, Peter Townend, Neil Whitaker and Mike Wilson.

ABOVE *Tornado* **passes Alexandra Palace with the 'Cathedrals Express' on 23 May 2009.** *(Robin Coombes/A1SLT)*

Chapter One

The history of the A1s

━━━◉━━━━━━━━━━━━━━━━

Sir Nigel Gresley, the most famous Chief Mechanical Engineer of the LNER, died in April 1941 while still in office, and the position was taken up by Edward Thompson. Gresley and Thompson had differing views on a number of key aspects of locomotive design. Thompson soon set about modifying a small number of the Gresley three-cylinder locomotives then in service, including one of the original 4-6-2 Pacifics and the six 2-8-2 Mikados of the P2 class.

OPPOSITE No. 60120 *Kittiwake* heads a northbound train in 1962. *(Cedric Clayson/A1SLT)*

The rebuilt Pacific, No. 4470 *Great Northern*, with a 6ft 8in wheel diameter, was the prototype for a new design of express passenger locomotive for the East Coast Main Line. This was re-designated A1, later to become Class A1/1, and renumbered in the new BR series as 60113.

Planning then started for this new class of locomotive during Thompson's tenure, but little progress was made by the design office, preference being given to the Class A2 6ft 2in-wheeled Pacific based on the rebuild of Gresley's Class P2, which was considered more suitable in war time and shortly after for the range of duties on the ECML.

Thompson retired in 1946 and was succeeded by Arthur Peppercorn, who was much closer to Gresley in terms of his locomotive design views. Production was stopped on the A2s at 15 locomotives, until the design had been modified to Peppercorn's requirements. This consisted principally of reverting to a cylinder arrangement where the outside cylinders were brought forward above the bogie to eliminate the lengthy exhaust steam pipework, which had led to problems of fracture in operation. A further 15 locomotives of this class were then built, of which *Blue Peter* was one.

More importantly, design work on the larger-wheeled A1 was stepped up, but now to Peppercorn's own designs. The intention was to create a powerful locomotive capable of hauling trains of up to 600 tons at 50mph, or 500 tons at 60mph, in line with the thoughts of the Board at the time, to increase capacity by running longer trains. Tests with a dynamometer car soon after their introduction showed that they were more than capable of achieving this requirement at economical rates of fuel and water consumption. In practice, as track was renewed and line speed was increased in the early 1950s, this thinking changed to a policy of running shorter but more frequent and faster trains, and the A1s were rarely tested to their full capacity.

Orders were placed for these locomotives in the latter months of the LNER as an independent company, but the nationalisation of the railways on 1 January 1948 meant that all were delivered in the British Railways' era. In total, 49 locomotives were built, with the work being shared between the Doncaster Plant and Darlington North Road Locomotive Works (26 at Doncaster, 23 at Darlington).

The locomotives were built with two-piece main frames. The reason for this was the restricted availability of steel plate of the required size so soon after the end of the war. The frames had a 4ft 8in overlap from the front bogie and over the leading coupled wheels, and the two sections were riveted together with driven, fitted rivets (cold pins). Later, there were some occurrences of frame cracking in the front sections, and strengthening of the frame in front of the overlap was eventually required.

BELOW Peppercorn A1 class 4-6-2 No. 60127 when new at the weigh-house in Doncaster Works in 1949. This was the first locomotive to be produced in the new blue livery. It later received the name *Wilson Worsdell*. *(Peter Townend/A1SLT)*

They were fitted with riveted steel boilers with a copper inner firebox, rated at 250psig working pressure. The fire grate area was 50ft^2, significantly greater than the Gresley A3s and A4s which had a grate area of 41.25ft^2. This had been designed to compensate for the lower quality coals available just after the war, but in practice, as coal quality improved, this change was found not to have been necessary. The driving wheel diameter was 6ft 8in (one of the major differences from the Thompson and Peppercorn A2s which had smaller driving wheels at 6ft 2in).

They had three cylinders, one internal (between the frames), and each was fitted with Walschaerts valve gear. Five (Nos 60153 to 60157) were fitted with Timken roller bearings on all the axles as part of a testing programme which led to the BR Standard designs, and in an attempt to increase life between overhauls. This was reported to be largely successful, but was not extended to the rest of the class. Kylchap exhausts were fitted, along with double chimneys. In later life, they were all fitted with the AWS (Automatic Warning System) which alerted the crew in advance to the aspect of a signal.

The locomotives weighed 104 tons 14cwt, the five with roller bearings weighing about one ton more, with the tenders adding a further 60 tons. Their tractive effort was rated at 37,397lb. As these locomotives effectively spanned the transition from the old LNER days

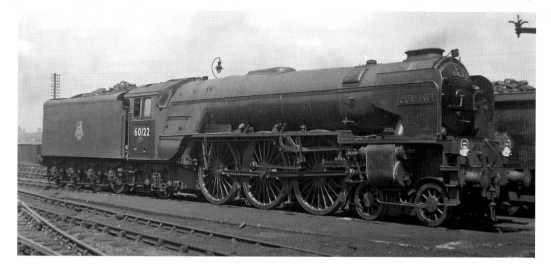

ABOVE Receiving attention from schoolboy trainspotters at King's Cross station in 1954 is No. 60157 *Great Eastern*. *(Geoff Parrish/ A1SLT)*

to the nationalised British Railways, they were seen in three liveries. When BR took over, they initially continued with orders already in the pipeline, and continued with the regional liveries. Later, a standard livery was introduced with blue being selected for main line express locomotives, dark green for mixed traffic, and black for freight locomotives.

Experience in use showed that the blue faded badly in operation, and this was subsequently dropped in favour of the dark green for express locomotives, often referred to today as Brunswick green, but more correctly, Middle Chrome Green.

Locomotives Nos 60114 to 60126 built at Doncaster, and Nos 60130 to 60152 built at Darlington (the full complement for that works) emerged in LNER green, with BRITISH RAILWAYS on the tender sides. Nos 60127 to 60129, and the final ten, Nos 60153 to

60162 (all built at Doncaster), emerged in BR blue. Over the period September 1949 to June 1951, the remaining 36 LNER green engines were gradually repainted in the blue. From August 1951, when the blue livery was discontinued, subsequent repaints were all in BR Brunswick green. The BR emblem of the time started to appear on the tender sides from the introduction of the blue livery, later being replaced by the BR crest.

The locomotives were all named, mostly following their first general overhaul, although No. 60114 *W. P. Allen* was named a couple of months after entering service. The names were chosen in recognition of the constituent companies of the former LNER and their locomotive engineers, bird names formerly carried on the A4s, characters from Sir Walter Scott's novels, and the LNER Pacific favourite, racehorses. The name *W. P. Allen* carried by

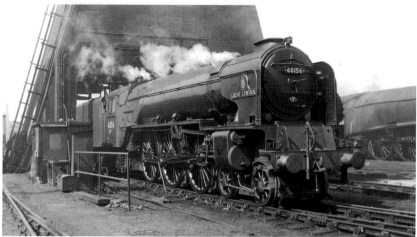

LEFT No. 60156 *Great Central* is seen under the coaling plant at King's Cross motive power depot (known as Top Shed), in 1957. Gresley A4 Pacific No. 60034 *Lord Faringdon* can be seen in the background. *(Peter Townend/A1SLT)*

LEFT Heading the 'Tees Tyne Pullman' in 1957, No. 60139 *Sea Eagle* exits Gasworks Tunnel after leaving King's Cross station. *(Peter Townend/A1SLT)*

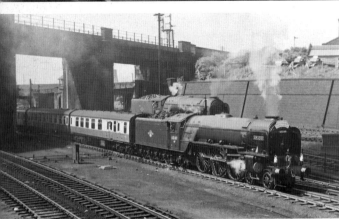

first Peppercorn A1, was that of a former Great Northern Railway employee who became the general secretary of ASLEF before joining the Railway Executive.

Most of their working lives were spent hauling the ECML expresses for which they had been designed, although it was not unusual, particularly in the latter years, to see them hauling heavy goods trains. They were allocated to main line sheds, primarily King's Cross, Grantham, Doncaster, Copley Hill, York, Gateshead, Heaton and Haymarket.

In the early 1950s, as train services improved after the war, BR reintroduced the prestigious summer-only non-stop service between London and Edinburgh with a gradually reduced journey time. This was latterly named the 'Elizabethan' and was hauled by Gresley A4s. For this service, the A1s were at a disadvantage to the A4s, which had corridor tenders, and allowed a crew change without stopping en route. The tenders were not interchangeable between the classes due to the A4s having vacuum braking, as opposed to the steam braking on the A1s.

The A1s were generally well received by both footplate crews and maintenance staff. They had a reputation for free steaming (the best of any of the Pacifics), being economical in terms of steam and water (in the hands of experienced crews), and for 'in-service' reliability. They were powerful engines, capable of handling any load for which they were allocated. On several occasions, the A1s were recorded at speeds in excess of 100mph down Stoke Bank (the scene of Gresley

LEFT No. 60117 *Bois Roussel* waits at King's Cross mpd, alongside Gresley A4 No. 60033 *Seagull* and A3 60108 *Gay Crusader* in 1959. *(Peter Townend/A1SLT)*

BELOW In less than pristine condition, Nos 60140 *Balmoral* and 60146 *Peregrine,* rest in a motive power roundhouse in 1964. *(John Arnott-Brown/ A1SLT)*

also insufficient and this could cause a side oscillation at speeds of 60mph, which made the engine ride uncomfortable. This was later improved by increased loading in the side springs, and as the track was simplified over the years the riding was improved considerably.

Their reliability led to some high mileages – over one six-month period the A1s allocated to King's Cross shed achieved an average of 6,400 miles per four-week period, compared with 4,800 miles per four-week period for the A4s. One roller-bearing A1 racked up 96,000 miles in a twelve-month period, the highest recorded for any locomotive.

Maintenance costs were exceptionally low. There were a number of incidents of injector failure, but investigation showed this to be due to problems with the water supply from the tender, rather than with the injectors themselves. There were also some incidents of big end failure on the driving rod from the inner cylinder, in common with a number of the three-cylinder engine classes, and the separate middle valve gear did fail on occasions, but generally the casualty rate was low. Lubrication of the inside cylinder components was difficult to access, and for this reason it is possible that it did not always receive the attention it required on a regular basis.

The locomotives regularly achieved mileages in excess of 100,000 between classified repairs, those with roller bearings averaging 118,000 miles.

In the early 1950s, BR carried out comparative tests on a selection of 7P/8P classified locomotives from the different regions with respect to maintenance costs, and the A1s came out well ahead – the comparative costs being:

ABOVE The end is nigh for No. 60129 *Guy Mannering*, minus nameplates, at Tyne Dock shed, shortly before being scrapped in 1965. *(Alan Willis/A1SLT)*

BELOW No. 60157 *Great Eastern* being cut up at Draper's yard, Hull, on 7 March 1965. *(Rev J. David Benson/A1SLT)*

A4 No. 4468 (later 60022) *Mallard*'s record-breaking run in 1938), although drivers of the time will say this was a daily occurrence.

The riding of the A1s was generally not found to be as good as the earlier Gresley Pacifics however. The design of the bogie was different with the weight taken on the side bearers instead of the centre as in the Gresley locomotives. This change, in combination with the Cartazzi axle at the rear, produced a ride which was sensitive to track irregularities such as points and crossings, which were more numerous on the railway at that time. The compression on the bogie side springs was

LNER	A1	8.53 pence per mile
GWR	'Castle'	9.73 pence per mile
SR	'West Country'	12.33 pence per mile
LMS	'Duchess'	12.70 pence per mile

The first withdrawal (of No. 60123 *H. A. Ivatt*, following an accident) came in October 1962, although others soon followed, and the process was completed in June 1966 when the last remaining engine, No. 60145 *Saint Mungo*, was withdrawn. Despite hopes that the final engine would be preserved, this was not to be, and they all succumbed to the cutter's torch.

The original Peppercorn A1 class 4-6-2 Pacifics

BR Number	Name	Built	Date to Traffic	Date Named	Date Withdrawn	Cut up at
60113*	Great Northern	Doncaster Plant	Sept 1945		Nov 1962	Doncaster Plant
60114	W. P. Allen	Doncaster Plant	Aug 1948	Oct 1948	Dec 1964	Hughes Bolckow, North Blyth
60115	Meg Merrilies	Doncaster Plant	Sept 1948	Jun 1950	Nov 1962	Doncaster Plant
60116	Hal o' the Wynd	Doncaster Plant	Oct 1948	May 1951	Jun 1965	Hughes Bolckow, North Blyth
60117	Bois Roussel	Doncaster Plant	Oct 1948	Jul 1950	Jun 1965	Clayton & Davie, Dunston-on-Tyne
60118	Archibald Sturrock	Doncaster Plant	Nov 1948	Jul 1950	Oct 1965	T. W. Ward, Beighton
60119	Patrick Stirling	Doncaster Plant	Nov 1948	Jul 1950	May 1964	Cox & Danks, Wadsley Bridge, Sheffield
60120	Kittiwake	Doncaster Plant	Dec 1948	May 1950	Jan 1964	Darlington Works
60121	Silurian	Doncaster Plant	Dec 1948	May 1950	Oct 1965	T. W. Ward, Killamarsh
60122	Curlew	Doncaster Plant	Dec 1948	Jul 1950	Dec 1962	Doncaster Plant
60123	H. A. Ivatt	Doncaster Plant	Feb 1949	Jul 1950	Oct 1962	Doncaster Plant
60124	Kenilworth	Doncaster Plant	Mar 1949	Aug 1950	Mar 1966	Drapers, Sculcoates, Hull
60125	Scottish Union	Doncaster Plant	Apr 1949	Jan 1951	Jul 1964	Cox & Danks, Wadsley Bridge, Sheffield
60126	Sir Vincent Raven	Doncaster Plant	Apr 1949	Aug 1950	Jan 1965	Drapers, Sculcoates, Hull
60127	Wilson Worsdell	Doncaster Plant	May 1949	Sept 1950	Jun 1965	Hughes Bolckow, North Blyth
60128	Bongrace	Doncaster Plant	May 1949	Nov 1950	Jan 1965	Drapers, Sculcoates, Hull
60129	Guy Mannering	Doncaster Plant	Jun 1949	Nov 1950	Oct 1965	R. A. King & Co, Norwich
60130	Kestrel	Darlington Works	Sep 1948	Jul 1950	Oct 1965	J. Cashmore, Great Bridge
60131	Osprey	Darlington Works	Oct 1948	Jun 1950	Oct 1965	T. W. Ward, Killamarsh
60132	Marmion	Darlington Works	Oct 1948	Dec 1950	Jun 1965	Hughes Bolckow, North Blyth
60133	Pommern	Darlington Works	Oct 1948	Apr 1950	Jun 1965	Clayton & Davie, Dunston on Tyne
60134	Foxhunter	Darlington Works	Nov 1948	Oct 1950	Oct 1965	T. W. Ward, Beighton
60135	Madge Wildfire	Darlington Works	Nov 1948	Oct 1950	Nov 1962	Doncaster Plant
60136	Alcazar	Darlington Works	Nov 1948	Dec 1950	May 1963	Doncaster Plant
60137	Redgauntlet	Darlington Works	Dec 1948	Jun 1950	Oct 1962	Doncaster Plant
60138	Boswell	Darlington Works	Dec 1948	Sept 1950	Oct 1965	T. W. Ward, Killamarsh
60139	Sea Eagle	Darlington Works	Dec 1948	May 1950	Jun 1964	Cox & Danks, Wadsley Bridge, Sheffield
60140	Balmoral	Darlington Works	Dec 1948	Jul 1950	Jan 1965	Drapers, Sculcoates, Hull
60141	Abbotsford	Darlington Works	Dec 1948	May 1950	Oct 1964	Drapers, Sculcoates, Hull
60142	Edward Fletcher	Darlington Works	Feb 1949	Oct 1950	Jun 1965	Hughes Bolckow, North Blyth
60143	Sir Walter Scott	Darlington Works	Feb 1949	Sept 1950	May 1964	Drapers, Sculcoates, Hull
60144	King's Courier	Darlington Works	Mar 1949	Jan 1951	Apr 1963	Doncaster Plant
60145	Saint Mungo	Darlington Works	Mar 1949	Aug 1950	Jun 1966	Drapers, Sculcoates, Hull
60146	Peregrine	Darlington Works	Apr 1949	Dec 1950	Oct 1965	T. W. Ward, Killamarsh
60147	North Eastern	Darlington Works	Apr 1949	Mar 1952	Aug 1964	Drapers, Sculcoates, Hull
60148	Aboyeur	Darlington Works	May 1949	Jan 1951	Jun 1965	Arnott & Young, Dinsdale
60149	Amadis	Darlington Works	May 1949	Dec 1950	Jun 1964	Cox & Danks, Wadsley Bridge, Sheffield
60150	Willbrook	Darlington Works	Jun 1949	Jan 1951	Oct 1964	Drapers, Sculcoates, Hull
60151	Midlothian	Darlington Works	Jun 1949	Mar 1951	Nov 1965	Station Steel, Wath Central
60152	Holyrood	Darlington Works	Jul 1949	Jun 1951	Jun 1965	J. Cashmore, Great Bridge
60153	Flamboyant	Doncaster Plant	Aug 1949	Aug 1950	Nov 1962	Doncaster Plant
60154	Bon Accord	Doncaster Plant	Sep 1949	Apr 1951	Oct 1965	T. W. Ward, Beighton
60155	Borderer	Doncaster Plant	Sep 1949	Mar 1951	Oct 1965	T. W. Ward, Killamarsh
60156	Great Central	Doncaster Plant	Oct 1949	Jul 1952	May 1965	Clayton & Davie, Dunston on Tyne
60157	Great Eastern	Doncaster Plant	Nov 1949	Nov 1951	Jan 1965	Drapers, Sculcoates, Hull
60158	Aberdonian	Doncaster Plant	Nov 1949	Mar 1951	Dec 1964	Hughes Bolckow, North Blyth
60159	Bonnie Dundee	Doncaster Plant	Nov 1949	July 1951	Oct 1963	Inverurie Works
60160	Auld Reekie	Doncaster Plant	Dec 1949	Mar 1951	Dec 1963	Darlington Works
60161	North British	Doncaster Plant	Dec 1949	Jun 1951	Oct 1963	Inverurie Works
60162	Saint Johnstoun	Doncaster Plant	Dec 1949	Aug 1951	Oct 1963	Inverurie Works

*The first of the class, No. 60113 Great Northern, was originally built as a Gresley A1 class in April 1922 – all the others in this class, which later became class A10, were rebuilt and re-designated as A3s, with the exception of 60113, which was rebuilt by Edward Thompson in 1945 and eventually designated as an A1/1.

Chapter Two

In the beginning...

By 1990, the steam preservation movement was well established in the UK with a wide variety of railway societies and preserved running lines around the UK, each tending to focus on locomotives indigenous to the region concerned.

OPPOSITE A section of the old Darlington Carriage Works, renamed Hopetown Works, was secured as a home for the build, and was opened on 25 September 1997. *(Peter Rodgers/A1SLT)*

Locomotives which had entered preservation with the demise of steam, along with wrecks still being recovered in various states of disrepair from Barry scrapyard, were still being gradually repaired and restored to service. Some of these required the manufacture of significant new parts to replace those missing or damaged.

The LNER express locomotive scene was represented by six Gresley A4 Pacifics (*Mallard, Bittern, Sir Nigel Gresley*, and *Union of South Africa* in the UK, *Dwight D. Eisenhower* in the USA, and *Dominion of Canada* in Canada), one Gresley A3 (*Flying Scotsman*), one Peppercorn A2 (*Blue Peter*) and one Gresley V2 (*Green Arrow*). The obvious omission from this list was the A1, and various LNER enthusiasts were wondering, individually, whether a new A1 could be built. Three of these were Mike Wilson (later to become the first Chairman of The A1 Steam Locomotive Trust), David Champion, and Ian Storey.

Mike Wilson's involvement started when plans were announced to redevelop Stockton station in the late 1980s, which would have led to the effective destruction of this historic site.

Along with others, he set out to prevent this happening, and to get the site listed status. As part of the brainstorming for this the idea was put forward that it would be an interesting project to construct a static steam locomotive for display in a bay platform – and the obvious choice for this, from the well-known locomotives of the LNER, was a Peppercorn A1.

Ian Storey was a steam enthusiast who had purchased 'Black 5' No. 44767, since named *George Stephenson*, and during a visit to Carnforth in connection with this in the 1980s he had noted an old boiler and a cylinder block, from A3 No. 60041 *Salmon Trout*. With his engineering expertise he started to think that it would be possible to build a new A1 – the boiler having been perceived as the major obstacle.

Through his business, Ian Storey met David Champion, a lifelong steam enthusiast, who had also been closely watching the steam preservation movement, and in particular, the items which were being remanufactured to repair heritage locomotives. Since the mid-1960s he had been convinced that one day, someone would start building steam locomotives again to replenish power on the growing number of heritage railways, once the elderly engines became too precious for everyday use. As he watched advance after advance in the heritage steam industry he realised that all the component parts for a new steam locomotive were now being made as part of different restoration projects. All that was needed was the general realisation that it was possible to build a new main line steam locomotive – and somebody to do something about it.

What was needed was a trigger to bring together the individuals who were having similar thoughts, and this occurred when Mike Wilson wrote to *Steam Railway News* with a proposal to set up a society to build a new A1. The initial response was favourable, and resulted in a meeting of a small number of those who had expressed interest, at York in March 1990. This culminated in a decision in principle, to go ahead with a new-build Peppercorn A1.

In April 1990, *Steam Railway News* carried the following article:

REALITY WITHIN 4 YEARS

-A **NEW** PEPPERCORN A1 IN STEAM

Time is of the essence
LETS MAKE AN A1 - YOU CAN MAKE IT HAPPEN

Contact Mike Wilson for details:
A1 Locomotive Trust Ltd, 24 Eleanor Place,
Stockton-on-Tees, Cleveland TS18 3JE.
Telephone: 0642 673241

RIGHT The first promotional poster displayed a rather over optimistic view of the project.
(Geoff Smith/A1SLT)

Peppercorn A1 Replica Steering Committee Formed – Parts and Drawings Sought

The planned construction of a replica LNER A1 Pacific has taken a step further, with the formation of a Steering Committee to get the project 'up and running'.

The committee have agreed to nominate the first one hundred members of the supporting group as life members, at a nominal cost of £5 each, to attract an initial hard core of members.

The Steering Committee is anxious to trace the whereabouts of any components or drawings for A1 Pacifics, a drawing, or drawings, for the cab being urgently required in view of the offer the Committee has received from an individual who is able to construct a cab for use as a publicity item and sales stand at events.

The Committee is also seeking members who have experience in the mechanical engineering, boiler making, and steel fabrication trades or who work in engineering drawing offices.

Anyone who is able to offer assistance concerning any of the above skills should contact Mr Michael Wilson at...

As a result of this article, David Champion contacted Mike Wilson, and also involved Ian Storey. A chance meeting also established that another business contact of David, Stuart Palmer (a lawyer), was also a steam enthusiast. David called him at his office, and said that he proposed to build a brand-new A1. Stuart asked: 'to what scale?' to which David replied '12 inches to the foot'. Over lunch that day, Stuart Palmer was brought 'on board' to assist with the legal structuring of the organisation. With the core group of four now together, discussions started on how best to move the project forward.

The earliest estimate of cost was £300,000, although this was quickly revised to £1 million. What was recognised was that this was the largest individual project to date within the steam railway preservation movement, and that the common model for the sector at the time of a 'society of enthusiasts' was not going to work

for a project of this size. What was needed was a professional organisation with experts in different fields, and a new approach to the issue of long-term fund raising.

What was needed was a business plan, and having given a lot of thought to the project, David Champion retired to his home office after dinner one night, and over a glass of red wine, in 20 minutes hand-wrote one and a half A4 pages which contained all the relevant elements

RIGHT Looking a little worn, the original hand-written 'business plan' put together by David Champion. *(Geoff Smith/A1SLT)*

of the plan. This was the 'eureka moment', although his better half was less enthusiastic when he subsequently announced that 'we are going to build an A1'. 'Yes dear' was the reply, as she continued with the ironing, not knowing that these words were to herald a complete change in their lives over the next ten years.

The two core elements of the plan were the financing, and the building of a business organisation with the range of expertise necessary to manage the project. Equally important, was the proposal that the locomotive would not be a replica of any of the original locomotives of the class, but simply the 50th A1. This allowed the locomotive to be built with the best modern manufacturing techniques and materials, and to be suited for the main lines of the 21st century.

The financing would be addressed by securing long term 'drip-feed' income from supporters. This had to be affordable, and at the time, the price of a pint of beer in the North East of England was around £1.25. It was considered that most supporters would be willing to sacrifice one pint per week, in order to contribute £5 per month to the project. The contributions had to be cheap to collect, and had to be collectable without the potential for the contributor to forget – hence monthly collection by standing order was considered to be the best method. Thus was born the 'Covenantor' – a crucial element in the business plan. One-off contributions would not be rejected, but these would be secondary, and

would not carry the same privileges as would a long-term covenantor.

UK tax law at the time gave tax benefits for regular donations given under a 'Deed of Covenant', which basically, was a commitment to continue paying donations for a minimum of four years. This was replaced in April 2000 by the Gift Aid scheme which continued to give a tax advantage for charitable donations, but without the need for the four-year commitment. Stuart Palmer was responsible for putting in place all the necessary paperwork for the birth of The A1 Steam Locomotive Trust, a registered charity, which allowed the Trust to take advantage of the tax rebates available under both schemes.

A further key element of the financing plan would be commercial sponsorship, which would normally take the form of materials and/or services at specially discounted rates – this was to prove crucial in the overall financing plan as costs rose.

Putting a professional team together with the range of expertise necessary to drive a project of this magnitude was equally important, and it was recognised that during the initial efforts to raise the profile of the project it would be essential to identify and recruit a competent team.

The first essential was to have a high-profile 'launch' meeting, and during the summer of 1990 efforts were made to get as much publicity as possible ahead of this meeting, which was to take place at the Railway Institute in York on 17 November 1990. The upstairs room, which had been chosen for the meeting, was packed out with people spilling into the corridor outside – some 120 potential supporters attending. It was clear that the project had captured the imagination of a generation of steam locomotive enthusiasts – this really was a 'first', and potentially a major boost for the industry.

Following the successful launch meeting it was decided to mount a series of high-profile 'roadshows' around the country with the dual aim of bringing on board as many covenantors as possible in a short time frame, and also to attract appropriate individuals to join the management team.

One of the first roadshows, in London, attracted three key members, who are still a

vital part of the Trust today – Mark Allatt (with his marketing expertise, and since 2000 has been the Trust's Chairman), David Elliott (with his engineering and project management expertise, and the Trust's current Engineering Director), and Barry Wilson (with his financial expertise, and the Trust's current Financial Director).

It was recognised, particularly in the early stages of the project where the number of covenantors was small, and hence the cash intake would build only slowly, keeping these people on board would be a very important element of the work. There were a lot of fairly high-profile cynics who were publicly stating that the project would 'never happen'. Indeed, many of these persisted with this attitude almost to project completion – and it was necessary to ensure that this message did not adversely influence those who had already signed up as covenantors, or those who may do so. Hence was born *The Pioneer,* a quarterly publication designed to show the progress of the project. Whilst this has changed name over the years, and is now known as *The Communication Cord,* it is still one of the benefits enjoyed by covenantors.

An annual covenantors' convention was also quickly established. This gave the opportunity

THE PIONEER
NUMBER 14   AUTUMN 1994
CONSTRUCTION COMMENCED
JOURNAL OF THE A1 STEAM LOCOMOTIVE TRUST

THE PIONEER
NUMBER 15   SPRING 1995
LOCO MAINFRAMES ERECTED
JOURNAL OF THE A1 STEAM LOCOMOTIVE TRUST

THE PIONEER
NUMBER 16   SUMMER 1995
ALL DRIVING WHEELS CAST
JOURNAL OF THE A1 STEAM LOCOMOTIVE TRUST

THE PIONEER
NUMBER 20   SUMMER 1996
EXTENDED 'NEW CYLINDER' ISSUE
JOURNAL OF THE A1 STEAM LOCOMOTIVE TRUST

for covenantors to meet, to hear first-hand from the management team what was happening, and to see the project at various stages of the build. These annual events continue to this day.

Continued recruitment of covenantors was vital, and every opportunity was taken to attend events of a rail-orientated nature to promote the project. In the early days, many events were attended with little or no return, but this was quickly refined, based on experience, and became more successful.

With the business structure now in place, together with the means to continue recruitment of, and communication with supporters, attention could turn to the primary objective – the construction of the locomotive.

The objective was to build a locomotive as close to the original Peppercorn design as possible, but accepting that some changes would need to be made to:

■ Meet current regulatory requirements for running on the main line network.
■ Take advantage of any developments in technology which would facilitate the build (bearing in mind that many components would be one-off), while keeping the locomotive's authenticity as an A1.

It was also decided that it would devalue the project if it was claimed that this was a direct rebuild of any individual member of the original class, and it should therefore follow the precedent of the North Eastern J72 class 0-6-0Ts, which were built by three successive administrations – NER, LNER and BR – over a 53-year period. The locomotive would simply be the next A1 and be given the next number in the A1 series, 60163.

One of the most significant potential early sponsors was New Cavendish Books, and while the committee were trying to decide on an appropriate name, Allen Levy – a world renowned expert on model locomotives and a board member of New Cavendish, suggested *Tornado*, in honour of the pilots who flew these planes in the first Gulf War. This seemed ideal – as well as honouring the pilots, the word also gives the impression of power – ideal for this locomotive.

While the engineers were now engaged in the practical issues associated with building the locomotive, intensive work still continued behind the scenes. Projected costs were rising, and there was a need to continue to attract and keep new covenantors. A further problem was also looming on the horizon – that of where the locomotive should be built. This needed a substantial building with heavy engineering facilities.

Doncaster, the birthplace of the original A1s, would have been ideal, and in July 1993 a presentation was made to the Doncaster Local Authority. This was enthusiastically received, and the council believed they could find a suitable location for the build. Meetings continued with the council, some potential properties were put forward and inspected, but were considered unsuitable, and the annual covenantors' convention was held in Doncaster in 1994. However, no further progress was made.

Worse was to come, when the first major parts, the frame plate blanks, were rolled, arrangements were made to have them profiled on the still-existing machinery at Doncaster Works. As the frame plates were loaded on to a truck to transport them to Doncaster, it was learned that, following a successful auction, the profiling machinery had been scrapped the previous day!

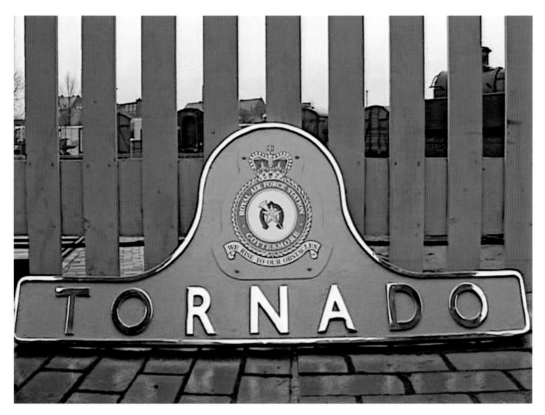

ABOVE The name is chosen. The *Tornado RAF Cottesmore* nameplate pictured at the time of handover by the RAF and representatives of other air forces at Tyseley. *(Ted Parker/A1SLT)*

By this time, Bob Meanley had been retained by the Trust for his project management skills, and with his connections he was able to offer to erect the frames at Tyseley Locomotive Works, by then in the hands of the preservation movement, but deep in Great Western territory. This offer was gratefully taken up and assembly of the frames, stretchers, cylinder castings and associated components began at Tyseley, under the auspices of Bob and his team.

However, being an LNER locomotive, the Trust was keen to see this back in what used to be the LNER area for the bulk of the build, and with construction now having started at Tyseley, this issue was becoming increasingly urgent.

In late 1994, an approach was made to Darlington Local Authority, and this time the enthusiasm was accompanied by positive action. Although the original locomotive works was no longer standing, having been demolished and replaced by a supermarket, the old carriage works – on the opposite side of the current North Road station, was available. An added advantage was that this was also adjacent to the Darlington Railway Museum, with whom a loose association could be established to mutual advantage.

The building needed substantial work before it would be usable, and with the assistance of the council, successful applications were made for grant assistance towards the work required. In total, a sum of £300,000 was provided by the European Regional Development Fund, the National Heritage Memorial Fund, and Darlington Council itself.

Work commenced on the refurbishment of the building in January 1997, and on 25 September the completed locomotive frames were moved to Darlington – and *Tornado* had its new home. This date was carefully chosen to reflect the fact that on the same day in 1825 the Stockton & Darlington Railway had officially been declared open.

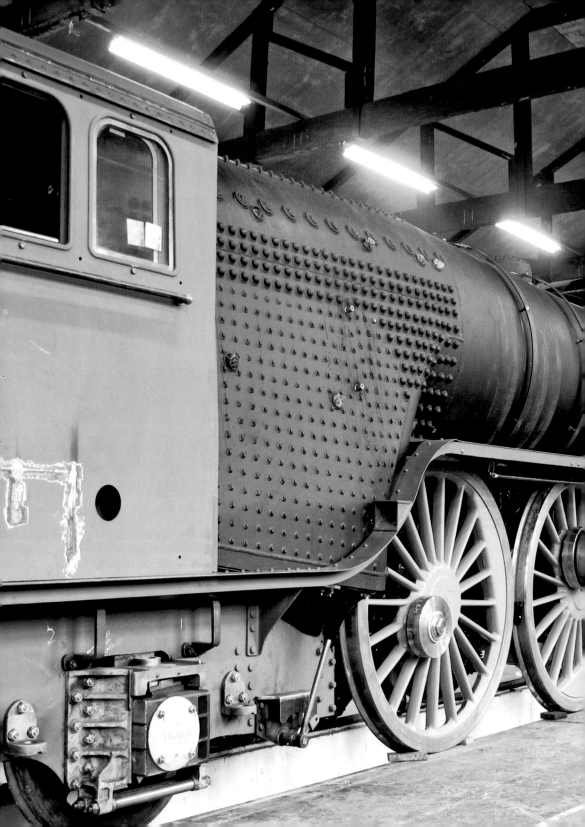

Chapter Three

Construction

━●━━━━━━━━━━━━━

When the original Peppercorn A1s were constructed in 1948/49, steam locomotive production was still in full swing, and the works were set up with full production facilities: pattern shops, foundry, fabrication, machining, assembly, paintshops etc. Locomotives were built in batches, and within a region there was a certain degree of parts commonality between different classes – all this meant that once designed, many components would have a significant production run.

OPPOSITE The boiler in place, clearly showing the crinoline rings, which are used to secure the cladding. *(Keith Drury/A1SLT)*

Building *Tornado* presented a different challenge. Although most of the drawings were available, each part built would now be a 'one-off', significantly increasing its cost, and requiring a lot of thought to be given to the best method of production, to minimise the time required for, and costs of making, a lot of special tooling. This included new patterns for castings, and jigs for assembly and testing.

Although most of the work would be carried out by paid contractors, with the assistance of volunteers where appropriate, and most would have relevant skills in the particular process involved, few, if any, would have been engaged in the actual construction of a steam locomotive.

A further problem was going to be the rate of build – *Tornado* was being funded by drip-feed covenantor donations, which would lead to a long build time (depending on the number of donors who came on board), and would need work planning which had to take account of funding availability. All these factors, plus the fact that this was a 'first', would combine to make the project a real, but exciting, challenge.

Design and drawings

Obviously, the first step for any major construction project is the design and drawings. As this project was to replicate the locomotives built in 1948/49, the first objective had to be to locate the original drawings. Enquiries were made at the National Railway Museum in York. It was known they had a lot of drawings, albeit these were in the archives and were in bundles of 40–50 as they had been received, and had never been sorted and catalogued. Exactly how many of these related to the Peppercorn A1s was uncertain, as was the degree to which these would form the requirements for a complete locomotive build.

The NRM was keen to protect its precious archives and extensive discussions took place as to how best to identify the relevant drawings, and obtain copies whilst ensuring these were preserved in their existing state. An agreement was reached with the museum under which the Trust would pay a significant sum of money, and would sort and catalogue the drawings, in return for copies of the any drawings related to

the Peppercorn A1s. This resulted in the Trust sending in a team of volunteers, supervised at all times by a member of the NRM staff.

It was estimated that the archive contained some 20,000 drawings from Doncaster, and thus identification of the relevant drawings proved to be a time-consuming task, made more difficult by the fact that the A1 class designation had been used three times for different locomotives, going back to the time of Nigel Gresley (including the Thompson and Peppercorn classes). Whilst dates helped to a degree, these were not conclusive, as the Peppercorn A1s used a number of parts in common with the earlier Gresley engines. This meant a lot of studying of the drawings, the compilation of a list, and a lot of cross-referencing. The arrangement drawings, where these were identified, contained a schedule of related component drawings – each of which had to be found, and some of which would contain further sub references.

After much hard work, some 1,100 drawings had been identified which were considered relevant, albeit that some were effectively duplicates (for example those related to the Timken roller-bearing locomotives as opposed to the plain white metal bearings of the bulk of the class). It was considered at the time that these represented over 90 per cent of the drawings which would be required during the build of the locomotive. The drawings were of various shapes and sizes, and of variable condition.

The next stage was to obtain copies of the originals, and consideration was given to the best method. Photocopying was rejected on two grounds – first because of the distortion which can easily occur during the copying process, and secondly, because of the inflexibility of the resulting copy. If further copies were required during the construction process these would become photocopies of photocopies, with consequent deterioration in quality, and also if modifications or additions were required to the drawings these would have to be completely redrawn.

Although digital scanning was in its infancy compared to the facilities available with today's powerful computers, a suitable digital scanner was identified, tested to demonstrate to the NRM staff that there was no risk to the originals,

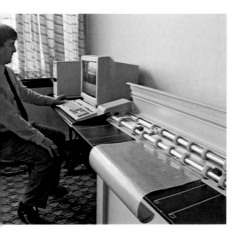

FAR LEFT The Trust's David Elliott operates the drawings scanner at the National Railway Museum at York in 1993, as the work to collect all the historical information necessary to facilitate the build is got underway. *(Ted Parker/A1SLT)*

LEFT More drawings await scanning and cataloguing at the National Railway Museum, York in 1993. *(Ted Parker/A1SLT)*

and then hired for an intensive period of scanning work.

Even this did not complete the process, as the scans picked up any dirt and marks on the originals, and therefore had to be digitally cleaned using specialist software. This was another time-consuming process and took place over a number of years as the drawings were required in the construction process. Initially, 140 drawings were cleaned to allow construction work to begin.

There were three further issues to be overcome. Most of the drawings did not show any machining limits and fits or tolerances. These would not have been so critical at the time, when most components would have been made in the same works, and conventions would have become custom and practice, or subject to local instructions. Also, if any problems were identified with tolerances during assembly it would have been a simple matter of returning the component to the machine shop for further adjustment. However, the building of *Tornado* would need many components to be made by a wide range of specialist sub-contract companies with the requisite skills, and then these would be brought together during assembly. It was therefore necessary to go through drawing by drawing and identify where mating components needed to have machining tolerances inserted.

The second issue was that there was often little or no information on the drawings relating to material specifications. The LNER had its own material specifications, issued to

LEFT Careful cataloguing and cross-referencing was an essential part of putting together all the necessary drawings. Alan Dodgson seen here at work. *(Ted Parker/A1SLT)*

LEFT The scanned drawings are stored on small computer tapes – the size of which is indicated by the key set placed adjacent. This was in the early days of personal computers, before the introduction of high-capacity storage devices. *(Ted Parker/A1SLT)*

the works but not shown on the drawings.

On the new build it was essential that current specifications for materials were used, to meet modern requirements and including those of the regulatory authorities.

As the ranks of the covenantors grew, they were joined by individuals who still worked in the railway industry, and had contacts in the appropriate departments. In particular, the BR Materials Engineering Section was able to provide information on material specifications for most of the major components. An additional major advance was the identification of a London Midland Region publication from 1957 entitled *Materials for Locomotives and Tenders* which gave a full specification for the materials to be used for each component. This document quoted BR specifications for materials, and had become the de facto standard for locomotive builds.

This was not quite the end of the story as, fortunately, the document also cross-referenced BR material specification to British Standard (BS) specifications. Whilst BS specifications have continued to evolve over the years, there is a continuous paper trail through to the present European Standards. Each specified material therefore had to be cross-matched to the current BS EN standard, for ordering purposes.

Finally, the drawings all showed sizes in imperial measurements. However, many materials are now only available in metric sizes. This required the dimensions to be converted to the nearest metric size. Although not a complicated task in itself, and usually resulting in only minor variations, the consequence of those variations had to be considered in detail where tolerances were close, to ensure that this would not result in complications on the operational locomotive.

To support the drawing work, and to provide some information which was not readily apparent from the drawings, physical measurements were taken of some components on A2 No. 60532 *Blue Peter*. Many of these were common with the Peppercorn A1, and with *Blue Peter* then in the Head of Steam – Darlington Railway Museum, this became a convenient source of some of the missing information.

Main frames and frame stretchers

The frames are the main structural members around which the rest of the locomotive is built, and are therefore the obvious starting point for any locomotive build. As stated earlier, the main frames on the original Peppercorn A1s were in two sections, joined by a lap joint which spans the horn cut outs of the leading pair of coupled wheels. This was a relatively unusual arrangement, most frames being in a single section, and the two-section arrangement did give some problems in service. A design review was undertaken, and it was decided that there was benefit in the main frames being in one piece, and this became the finally adopted arrangement. This did require the minor redesign of certain other components which fitted between the front end of the frames, to accommodate the decrease in frame width at this point (the front section of the original design was riveted outside the main section).

Behind the rear set of coupled wheels the main frames have an offset inwards to clear the backs of the trailing (Cartazzi) carrying wheels. Additional frame sections are then bolted on the outside of the main frames from the point at which they start to narrow, through to the rear of the locomotive. Short strengthening plates are inserted between the two frame sections at the point of the join to reinforce the frames at the point where their vertical height is at its

minimum, to clear the front of the firebox. The rear outer frames support the Cartazzi axle box horn blocks and the cab footsteps.

The original Peppercorn A1 frames were made out of $1^1/_8$in steel, the nearest modern equivalent being 30mm, which gave a small increase in thickness, and consequent strength. The frame plates were rolled in the plate mills at the Scunthorpe Works of British Steel on 22 April 1994 – a red letter day in the diary of the Trust.

The next stage was to have these profiled, and as mentioned earlier, it had been hoped to have this done at the Doncaster Plant where many of the original A1s were built. Sadly, this was not to be, as the machine was sold for scrap shortly before the plates were due to arrive. The work was therefore carried out at the British Steel Distribution (BSD) Plate and Profile Works in Leeds on their CNC Plasma and Oxy fuel profile-cutting machine.

The work involved cutting to shape, including cut outs for the horn blocks (where the coupled wheel axles pass through the frames). At this point the frames were moved to the works of TM Engineers at Kingswinford, where the edges were machined, all holes were drilled, and the offsets were made behind the rear set of driving wheels. This was carried out with the metal heated locally, adjacent to the bend, and forming the radius of the bend around a piece of large-diameter steel pipe temporarily bolted to the floor. Temporary stretcher rods were

ABOVE The plate for *Tornado*'s main frames was rolled in the plate mill at British Steel, Scunthorpe on 22 April 1994. *(Ted Parker/A1SLT)*

LEFT The side plates were laid out ready for profiling for *Tornado*'s frames at BSD, Leeds, on 13 July 1994. *(Ted Parker/A1SLT)*

FAR LEFT With David Champion looking on, the Trust's Honorary President, Dorothy Mather (widow of the late Arthur Peppercorn – designer of the original A1s) starts the cutting machine which will profile the frames, at BSD, Leeds. *(Ted Parker/A1SLT)*

LEFT Cutting (profiling) the side plates of *Tornado*'s frames in progress at BSD, Leeds. *(Ted Parker/A1SLT)*

bolted between the frames to allow these to be moved as an assembled item.

Although at this stage, the Trust was still in negotiations for the locomotive build to take place on what used to be LNER territory, there was no prospect of an early solution, and the offer to have the frames assembled at Tyseley Locomotive Works was gratefully accepted.

This involved fixing the frame stays – the sections which effectively hold the frames together (with most performing additional functions such as supporting the boiler), the horn blocks (which provide the guides for the sliding axle boxes of the coupled instead of driving wheels), as well as various attachment brackets to the outside of the frames to support the motion, footplate etc.

Working along the frames from the front, the first 'stay' is the front bufferbeam, mounted to the frames by means of gusseted angle brackets fixed to both sides of each of the frame plates. Behind this, and fixed at the bottom of the frames, is the stay known as the bogie top centre which supports the pintle, which locates the bogie itself. Above this is the inside cylinder casting, which incorporates the smokebox saddle (on which the smokebox is fixed). The

next mounting is the inside motion plate which provides the support for the motion from the inside cylinder – then comes the front boiler support stay mounted to the upper part of the frame, and the combined spring hanger brackets mounted to the lower part of the frames.

These are followed by a simple stay bent up out of plate which supports the reversing balance spring mechanism. Next is a large cruciform fabrication known as the star stay (a casting on the original locomotives) which carries the loco brake cylinder and rigging and which has been modified to accept two air-brake cylinders instead of a single steam cylinder, and to carry one of the two steam-driven air pumps. The top of the star stay also forms the rear boiler support. To the rear of this are the front firebox support brackets, with a cast frame stay between them. Two rear firebox supports are bolted between the inner and outer rear frame plates. Finally, there is the dragbox, to which the tender drawbar is attached.

Other components fitted while at Tyseley included the cylinder castings (see next section), the horn blocks, and the support brackets for the running boards.

Whilst many of the attachments on the original A1 were riveted, a decision was made that on *Tornado* they would all be bolted, using driven or fitted bolts retained by self-locking nuts. This avoided riveting which has become much less common as the heavy construction industry has declined in the UK, and also facilitated easier removal at a later date should

ABOVE The completed frames, with stretchers and other attachments are ready to leave Tyseley. *(Ted Parker/A1SLT)*

LEFT En route from Tyseley the completed frames, with centre cylinder casting in position, were displayed for a while in the Great Hall at the National Railway Museum at York. *(Rob Morland/A1SLT)*

RIGHT The frames, now with the basic cab attached, arrived outside Hopetown Works at Darlington on 25 September 1997. *(Peter Rodgers/A1SLT)*

RIGHT A short time later and the assembly was positioned inside Hopetown Works. *(Peter Rodgers/A1SLT)*

BELOW The three large cylinder castings were displayed outside Tyseley Works during the covenantor visit in 1996. *(John Rawlinson/A1SLT)*

this ever be required. An additional factor which reinforced the decision was concern about the noise which would be generated by riveting, both for the people involved in the work, and for neighbouring properties.

Cylinders, valve chests, pistons and valves

The cylinders are the point at which the heat energy in the steam is converted to mechanical energy to drive the wheels, and are the most complex castings on the locomotive. The A1s have three cylinders, the two outer cylinders which are clearly visible, and a third inner cylinder which sits between the frames above the front bogie.

The cylinder castings are very complex in that they contain two main chambers – one housing the piston (which is moved backwards and forwards by steam pressure), and the second housing the valves which control the admission of live steam to the cylinder, and allow the exit of the exhaust steam. The pistons on a steam locomotive are double acting, which means that they are moved in both directions by steam pressure. This means that the steam has to be admitted alternately from each end of the cylinder, while the spent steam is simultaneously exhausted from the opposite end.

The patterns for the cylinders were made by Kingsheath Patterns and were very complex. The pattern for the inside cylinder included no fewer than 28 cores which form the open spaces in the casting (and filled the load space of a Luton van). This casting includes, on its upper side, the smokebox saddle, which supports the weight of the smokebox. Whilst the outer cylinders had fewer cores, they too were complex castings.

The inside cylinder is bolted directly between the main frames and therefore the width of the casting had to be slightly adjusted from that of the original Peppercorn A1s, due to the difference in width between the main frames at this point (resulting from the use of single-piece frames). The outside cylinders are bolted directly to the outside of the main frames.

A decision was made that the cylinder castings would be in spheroidal graphite (SG) cast iron, as opposed to the grey cast iron used on the originals. This grade of cast iron is more ductile and has greater strength, and was therefore considered less likely to suffer from stress cracking over time. However, this did introduce one small problem – in grey cast iron the widely dispersed carbon particles within the metal act as a lubricant in operation, whereas in spheroidal cast iron the carbon is present in the form of more concentrated nodules which do not have any significant lubricant effect. To overcome this, grey cast iron cylinder and valve chest liners were fitted.

The pistons are forged in one piece with the piston rods, which run parallel to the centre line of the cylinder, through sealing glands in the integral rear cover, to the crosshead. The pistons are fitted with high-grade cast iron rings, running in the cast iron liner. The cylinder front end covers are SG iron castings, dished to match the shape of the domed piston. These are designed to minimise the free space in the cylinder when the piston is at the end of its travel, space which is occupied by steam that provides no useful work. By minimising the space, increased efficiency of steam use is gained.

The valve chest is parallel to the cylinder and is fitted with two piston valve heads on a common valve spindle. Ports connect each end of the cylinder to the valve chest. Where the ports emerge into the valve chest grey cast iron liners are fitted. The liners correspond to the range of travel of the respective piston valves. Each valve is fitted with four high-grade cast iron rings, running against the liner. Although the casting has a single port to the cylinder at either end, in order to provide adequate strength in the liner and to support the valve rings, the liner has a total of ten trapezium-shaped holes with diagonal webs between them. Steam from the boiler is supplied to the space between the two valve heads, and exhaust is ducted away from the outer ends of the valve chest. As the valve spindle moves fore and aft, the valve heads alternately allow live steam into the cylinder and exhaust to escape from the cylinder at each end.

Although the valves are specified as 10in diameter, the rear valve is actually $9\,^7/_8$in diameter. This is to ease removal of the valves and valve spindle for inspection through the front of the valve chest. The valve spindle passes through the rear valve chest via a gland to minimise steam leakage and the front of the valve spindle is supported in a bronze bush pressed into the valve chest front cover.

ABOVE The cylinder
as viewed on the
completed locomotive,
showing the cladding,
and the positions of
the drain cocks (one
at each end of the
cylinder), and the
safety valve (below the
main piston rod). (Geoff
Smith/A1SLT)

Wheels, axles and tyres

All wheels on the locomotive and tender are in cast steel, fitted with separate tyres. There are 20 wheels in total, the six main driving (coupled) wheels, four front bogie wheels, two rear wheels on the Cartazzi axle, and eight wheels on the tender. All were cast, along with most of the rest of the steel castings on the locomotive and tender, by principal sponsor, William Cook Cast Products. The casting process described here is the same, in principle, for all the steel castings on the locomotive.

The starting point for any casting is the design – in most instances the Trust was able to supply the drawings from the original Peppercorn A1 wheels. These then had to be critically examined by William Cook's design team to ensure that the resultant casting would have the required strength at all critical points, particularly around the centre boss, and where the spokes join both the boss and the outer rim. Whilst the sizes varied the principles are the same for all the wheels.

Once satisfied that the basic casting would achieve the desired aim, Cook's team then had to consider how the casting could physically be made. This involves deciding where the metal will be poured into the eventual mould, and how it will be distributed around the mould, using runners (feeder channels) where appropriate. Part of this process involves running a test casting in a three-dimensional computer modelling environment. This looks at the anticipated cooling characteristics at different parts of the casting as the metal is poured, to ensure the metal will flow to all parts, and will set as a solid mass, without cracks occurring as a result of varying cooling rates as the metal flows around the mould.

With the design work complete, the next step is the one of making a wooden pattern for each size of wheel. The patterns are made from tulipwood, chosen for its easy workability, while at the same time being fast growing and therefore considered sustainable. The patterns are made in sections which are then assembled together. Once made, the patterns can be used over and over, hence the need for only one pattern for each size of wheel. In the case of the large coupled wheels, the pattern was

As well as the passages for the steam flow – inlet, through the valve gear, through the cylinder, and to exhaust – the castings also have to have provision for other fittings, most notably the cylinder drain cocks and pressure relief valves.

When starting up from cold or after a period of standing idle, the steam in the cylinders will condense to water. Water, unlike steam, is incompressible, and unless this is released the movement of the piston will cause a massive build-up of pressure and result in major damage to the piston, the cylinder casting, or both.

To deal with this the cylinders incorporate two safety features. The first of these is the cylinder drain cocks, which are manually controlled valves fitted at the bottom of each end of each of the cylinders (six in total). These are joined together by a series of levers and pull rods so as to open together and close together. The operating mechanism passes from the inside to the outside of the frames, and then along the side of the locomotive to a lever in the cab. This allows the locomotive crew to open the drain cocks prior to starting to move the locomotive, characterised by the clouds of steam which are experienced emerging from the front of a locomotive immediately prior to movement.

A secondary safety feature is provided by pressure relief valves fitted to the cylinders (and adjusted to slightly above normal working pressure) and which would open in an emergency to release excessive pressure before any damage could be caused to the cylinder casting itself.

made in a modular fashion, permitting small changes to be made in the pattern for the boss and counter balance weights for the different wheels. Pattern making is a complex and expensive process, and each wheel pattern can take up to four weeks to make.

The pattern is then transferred to the casting shop where the moulds are prepared – the mould consists of two parts, the lower part (known as the drag), and the upper part (known as the cope). Air-setting sand (special sand with a resin additive) is built up around the pattern in the drag, in such a way that when the sand is set the pattern itself can be removed. The process is repeated in the cope so that when the two halves are brought together, the mould forms the hollow shape made by the pattern. Also added at this point, are the runners (which will take the metal to the different parts of the casting), the risers (which allow the air to escape as the metal fills the voids), and the pouring cups through which the hot metal will be poured. Once both halves of the mould are prepared, they are painted internally with a special paint which reduces the chances of the hot metal penetrating into the sand itself.

Basic steel plate is melted in an electric induction furnace and, dependent upon the specification for the casting, various additives may be incorporated into the molten metal. Prior to use the metal is sampled and laboratory tested to ensure that it is within specification. It is then poured into the mould and allowed to solidify – depending on the metal specification the mould may be removed while the metal is still hot (but solid), or it may be left in the mould to cool fully before mould removal.

The casting, as it leaves the mould, bears little resemblance to the finished item, with all the runners, risers etc. still attached, so the next stage is to clean up the casting. A combination of sand blasting, torch cutting and careful grinding is necessary to achieve this.

The next step is to heat treat the casting – this is placed in a furnace and brought up to a specified temperature depending on the type of treatment required, held for a specified period of time, and then either air cooled, or 'quenched' (cooled rapidly by immersion in a bath of water or oil). The usual requirement for this is to even out the stresses in the metal caused by the casting process, although heat treatment can also be used to change the metal characteristics (e.g. surface hardening).

The casting then goes through a process of NDT (non-destructive testing). This starts with a visual inspection, looking for any flaws, followed by ultrasonic testing (which will detect any internal cracks), and is finally Radiographed in the most critical areas. The testing process is prescriptive in order to comply with the requirements of an external standard, in this case the Railway Group Standards. Certificates of compliance will accompany each casting, for future reference.

Any flaws uncovered during the testing process then have to be rectified, by grinding or burning out the flaw, filling with weld, grinding down and re-testing to ensure that the problem has been completely eliminated. At this point, the casting can go into the machine shop,

BELOW The Trust's David Elliott (right) and pattern-maker Mick Stockdale, from William Cook & Sons, stand in front of the driving (coupled) wheel pattern. *(Phil Champion/A1SLT)*

where it receives primary machining to bring it close to the drawing dimensions (while leaving some excess metal for final machining during assembly and set-up by the customer). In the case of the wheels this included initial boring out of the hole in the centre of the boss to take the axle (the boss was cast solid to avoid causing any weak points in the boss due to the metal flow during the casting process). Subsequent boring of the hole required a high degree of accuracy to ensure exact centring.

At this point the wheel centres are now ready for delivery.

The locomotive axles were provided by Firth Rixon Forgings. The process of forging involves starting with an oversize section of metal, which is heated and then formed using hydraulically operated hammers, to bring it to shape, and close to the required size (similar to Blacksmithing, but on a much larger scale). On completion of the forging process, the section is then subject to preliminary machining, heat treatment and again subject to non-destructive testing to ensure integrity prior to supply to the customer.

Each of the wheels is fitted with a 'tyre', which is the wearing surface in contact with the track. The idea of the tyres, similar to road vehicles, is that when these wear down over a period of time they can be replaced without the need to cast a completely new wheel. As manufacturing processes have evolved over the years, simplifying the making of wheels, and the wheels on modern rail vehicles have become smaller, most wheels are now of monobloc construction and when these are worn down they are simply replaced as a single unit. This has meant that tyre manufacturing capability has disappeared from the UK, and the tyres for *Tornado*'s wheels had to be sourced from Belgium and South Africa.

On the original A1s the tyres were riveted via a lip on the front surface of the tyres, through the wheel centre casting. The risk of the associated holes weakening the tyre and wheel centre, which could lead to fractures, has led to riveted tyres no longer being acceptable for new-build locomotives under the Railway Group Standards. The method of fixing which has been adopted, is known as the 'double-snip' arrangement. With this method, a small groove is machined into the rim of the wheel at the rear.

The tyre has a corresponding lip on its inner surface. The tyre then has to be heated and expanded before it is shrunk on to the wheel centre, the lip then prevents any sideways movement of the tyre.

The wheelsets were all assembled at the Bury workshops of Riley & Son (E) Ltd (who have the necessary railway authority approvals to carry out the process). This is a critical part of the overall construction of the locomotive – the wheels clearly have to be a tight fit on the axles, they must run true and be well balanced (to avoid excessive wear on the track), and the tyres must be correctly fitted to prevent any possibility of separation during operation.

The general process was the same for all wheelsets, the three main coupled pairs, three bogie sets (two front, one rear), and four tender sets, although the process on the main coupled sets was by far the most complex.

The process starts with the final machining of the axles to the correct dimensions, together with the final machining of the wheel centre bosses. At this time one wheel in each set is also machined to final dimensions. The axle ends, where the wheel is to be fitted, are machined with a 1-in-500 taper, whilst the hole in the centre boss of the wheel is machined parallel.

Prior to fitting the wheels, the Timken roller bearings had to be fitted onto the axles – these are complete units, with no facility for splitting the outer races. This is essential to achieve the close tolerances required. The bearings themselves are twin-taper roller, allowing for a small degree of side thrust, as well as the normal loads associated with the rotational movement of the axle. Whilst the original Peppercorn A1 bearings were oil lubricated, *Tornado*'s bearings are grease lubricated, making it easier to avoid accumulation of moisture in the bearings, which would lead to premature corrosion and the need for changing. This is very important as replacement of the bearing would involve removal of the wheels themselves – not an easy task. The bearings are rated for one million miles – or 67 years at 15,000 miles per year!

The wheels are then cold pressed on to the axle, using a 400-ton press. (Normal practice is for the wheels to be heated and shrunk over the axle, but this method could not be used with

Tornado due to the use of the roller bearings, which were fitted to the axles before the wheels, and which could potentially suffer damage at the high temperatures involved). In this process, it has to be ensured that the wheels are set on the axle so that the crank pins are at the correct relative angle.

It is important that the pressing force applied during the fitting is correct – too little and there is a danger of the wheel working loose, too much and the metal is physically deformed, which would potentially cause stress failure in operation. To ensure that the correct force was used, a series of tests were carried out on a dummy wheel centre boss and scrap axle, measuring the performance of the metal over a range of applied pressing force. The appropriate setting was determined and monitored on the press during the actual fitting of the wheels. To assist the fitting, lubrication was in the form of rape seed oil, obtained from a local branch of Tesco (not everything in the build was high tech!)

With the wheels pressed into position, the final machining of the part-machined wheel rims in each set is carried out. The tyres are then fitted to the wheels by heating the tyre, fitting it over the wheel and cooling so that it is shrunk into place. The tyres are then machined to the final dimensions and tread profile.

The final requirement is to balance the wheelset, similar to the balancing of a car tyre. The wheelsets are spun at speed to identify the point on the circumference where weight needs to be removed. On the non-coupled sets it is simply a case of grinding off small amounts of

ABOVE Not everything is high tech! Rape seed oil was used as lubrication during wheel pressing on the rear (Cartazzi) axle at Riley & Son (Electromech) Ltd of Bury. *(David Elliott/A1SLT)*

LEFT The coupled wheelsets stand after wheel pressing at Riley & Son (Electromech) Ltd, Bury. Note the protected wheel bearings fitted to the axles, prior to the pressing of the wheels themselves.
(David Elliott/A1SLT)

FAR LEFT The front coupled wheelset after the pressing of the wheels, showing the split axle arrangement to accommodate the drive from the centre cylinder piston.
(David Elliott/A1SLT)

LEFT The smaller front bogie wheelsets after wheel pressing.
(David Elliott/A1SLT)

metal, on the inside of the wheel so as not to affect its appearance, and retesting until the set is in balance.

Whilst the general principles described above also applied to the coupled wheel sets, there are a number of additional complications. The first relates to the axles themselves – whilst the centre and rear sets are on a solid axle, the front set is on a multi-part axle, to accommodate the drive from the centre cylinder. The concept is similar to the design of a car crankshaft, with the crank part of the axle (connected to the piston rod) offset from the centre line of the axle so that the movement of the piston creates a circular motion in the axle itself. Each of the wheels is fitted with a shorter bearing axle, which in turn is connected into crank webs. The other end of the web forms the centre line for the crank pin.

To assemble this structure the webs are heated and shrunk on to the inner ends of the stub axles, then the crank pin is shrunk into the holes in the ends of the webs. As it is difficult to maintain angular accuracy of the crank pin and webs small errors are corrected by leaving one of the stub axles oversize so that it can be machined to the finished size to true up the whole crank axle assembly. To reduce the angular assembly error to a minimum a special jig had to be prepared to assemble the crank axle. The wheels were then fitted to the crank axle in the same way as the other wheelsets.

All the coupled wheel sets are provided with keys between the wheel hub and the axle as an additional safeguard against movement of the wheel on the axle. Movement of wheels on axles

for non-coupled wheelsets is not likely as the rotational forces are not great, however for coupled wheelsets the rotational forces are considerable, and this makes use of keys essential.

A further critical issue with the coupled wheelsets was to ensure the correct positioning of the crank pins. Those on the centre coupled wheels take the drive from the outer cylinders, and the one on the front set takes the drive from the centre cylinder. The crank pins on the centre set must be offset by 120° side to side, with the centre crankpin offset by 240°, to even out the power over each cycle of the pistons. With these set, the outer pins on the front and rear driving wheelsets must be aligned exactly with the centre pins on the respective side to take the connecting rods. Bearing in mind that each of these wheelsets weighs some 5.5 tonnes, it is clear that setting these up correctly was no mean feat!

The positioning of the crank pins on the wheels was achieved by accurately boring the crank pin holes using a large horizontal borer with digital readout to set the centres of the holes precisely. The crank pins were then fitted into holes in the wheel bosses by cooling the pins in liquid nitrogen, fitting them into the hole and allowing them to expand as they warmed up, creating a tight fit in the hole. They are additionally welded to the backs of the crank pin bosses for security. The inside cylinder crankpin forms an integral part of the leading coupled axle, as described above, and the connecting rod for this axle has a big end bearing with a split and bolted cap (again identical in concept to the internal combustion engine piston rod connecting to the crankshaft).

Although Riley's at Bury were able to carry out most of the work on the wheelsets, their lathe is only capable of working on wheel diameters up to 6ft 0in. This meant that the driving wheelsets, once the wheels were fitted, had to be taken down to the Severn Valley Railway workshops for final machining to size, sent back to Bury for the fitting of the tyres, and then back to the SVR for final machining of the tyre profiles. By the time these wheelsets turned for real on *Tornado*, they were already well travelled!

The final process with the coupled wheelsets was that of balancing. Whereas with the bogie and tender wheelsets this was simply a matter of dynamic balancing (similar to car wheels), with the driving sets the need is rather different. As well as the rotational forces which are present in all wheels, the coupled wheels are also subject to reciprocating forces – the forces imposed by the piston travelling backwards and forwards.

Trying to balance the coupled wheelsets is therefore very complicated, and involved a lot of theoretical work being carried out by a highly experienced specialist in this area. Each of the wheel castings contains a solid segment which is the balance weight, and balancing is created by drilling out face metal from this weight as required The objective is to fully balance the rotating masses (coupling rods, part of the connecting rods) and a proportion of the reciprocating masses (pistons, piston rods, crossheads, part of connecting rods and valve gear). More reciprocating balance makes for a better riding locomotive, however the extra weight added to balance reciprocating masses has nothing to do when the balance weights are in the top and

bottom positions as the wheels rotate. This gives rise to a vertical cyclic force on the track known as hammer blow, so it is necessary to restrict the reciprocating mass balance to a level which limits the hammer blow to acceptable values.

The complex calculations carried out produced a paper pattern, pre-marked with the drill points and sizes for each of the wheels. This was fastened to the back of the balance weight on each wheel and permitted accurate drilling out of the required blind holes to remove the specific weight of metal. On the original Peppercorn A1s the holes would have been drilled while the axle was on the balancing machine. In the case of *Tornado,* this was not practical (the machine used was designed for large electric motors), so fine adjustment was generated by drilling a small number of extra holes and achieving the required balance by inserting steel rods into some of the holes – this is similar to the process of adding clip-on balance weights to a car wheel to achieve dynamic balance.

ABOVE LEFT One of *Tornado*'s rear axle tyres is machined prior to fitting to the wheel at Riley & Son (Electromech) Ltd of Bury. *(David Elliott/A1SLT)*

ABOVE With tyres fitted, the Cartazzi axle wheelset was now complete. *(David Elliott/A1SLT)*

LEFT Back at Hopetown Works, the frames were positioned over the completed wheelsets. *(Fastline Photographic/A1SLT)*

With the wheelsets assembled and balanced, these then had to be fitted into the frames. For the middle and rear coupled wheelsets this involved the use of 'cannon' boxes. These are cast steel boxes, split to go around the axle, and bolted together to form a sealed housing. The ends of the castings are machined and fitted with 11 to 14 per cent manganese steel liners which slide in the horn blocks that are in turn, bolted to the cut outs in the main frames. Manganese steel has a property of work hardening, so as the locomotive is used the liner hardens to form a very durable wearing surface. Obviously, the leading coupled axle could not be fitted with a full-width cannon box as this would interfere with the drive from the centre cylinder, and two shorter sealed boxes were used, one on each stub axle. Once fitted up into the frames, solid bars (known as horn stays) are bolted to the frames underneath the horn blocks to prevent spreading of the horn block and also to prevent the wheelsets dropping out if the locomotive is lifted.

Motion

The motion is the means by which the movement of the pistons is transferred to the driving wheels, while at the same time controlling the movement of the steam valves to admit the steam at the correct points in the cycle. Bearing in mind that when travelling at 75mph the driving wheels of the locomotive are rotating 315 times per minute (more than five times per second), and each rotation represents one complete cycle of the piston and valve gear, it is clear that this is a critical area of the construction. Even minor errors would either lead to major inefficiency in steam use, or in the worst case scenario, would overstress the components of the motion and could potentially lead to catastrophic failure.

The first element of the motion is the means by which the linear movement of the piston is transferred to a circular movement of the main driving wheels. The piston rod passes through a pair of non-ferrous metal glands as it exits the cylinder to give a steam-tight seal. The piston rod is connected to the crosshead, which in turn runs in guides (the slide bars) which must be set perfectly parallel to the cylinder. The crosshead is also connected to the small end of the connecting rod, the other end being connected to the centre coupled wheelset by means of the crankpin. The crankpin is offset from the centre of the wheel, thus the backwards and forwards motion of the crosshead means that the connecting rod induces a circular movement at the wheel.

The three sets of coupled wheels are connected together by coupling rods, which allow the power from the connecting rod to be spread across the three wheelsets. The crankpins on the wheels on each side of the locomotive are set identically, but are offset from side to side by 120° to even out the thrust from the three cylinders. The inner cylinder is the same in concept, but on this instance, the connecting rod drives directly onto the crankshaft of the leading axle.

The second element of the motion is the valve gear, which controls the timing and quantity of steam admitted to the cylinders to create the piston movement. The A1s, in common with most later express locomotives in the UK (and around the world), were fitted with Walschaerts valve gear. This was developed by a Belgian mechanical engineer, Egide Walschaerts, in 1844. This is designed to allow variable 'cut off'.

In a simple double-acting piston arrangement, steam would be admitted through the valve to one side of the piston to push it back almost to the end of its travel, before the steam is cut off, and is admitted to the opposite side of the piston to push it in the opposite direction. This gives maximum power from the piston, but is very wasteful in terms of heat energy, since the steam does not have the opportunity to expand fully before the exhaust stroke is started. On a steam locomotive it is desirable to have a wide range of cut off adjustment to suit different running conditions and allow the driver to minimise the use of steam. When setting off, and maximum power is required, the cut off would be as late in the travel of the piston as possible – usually around 75 per cent of its travel.

Once the locomotive is up to speed cut-off can be much earlier in the travel, giving the steam time to expand fully before it is exhausted. The point of cut off is referred to in percentage terms as the proportion of the travel of the piston before the steam admission valve is closed (i.e. 25 per cent cut off would mean the piston had travelled 25 per cent of its distance before the steam supply is shut off). Walschaerts valve gear provides this facility, as well as the facility for 'reverse' operation where the steam is admitted to the back of the piston first, causing the motion to be in the reverse direction.

The valve gear consists of a range of rods, levers, and pivot arrangements as shown in the drawing overleaf.

From the crankpin of the main driving wheels an eccentric crank links to the eccentric rod, connected to one end of the radius (or expansion) link. The radius link pivots on a bracket attached to the main frames, and is slotted, allowing the radius rod to slide up and down within it. The rear end of the radius rod is supported by a die block in the reversing shaft arms. The reverser in the cab is operated by the driver. On turning in one direction, this causes the radius rod to fall in the radius link, putting the motion into forward gear. By turning in the opposite direction, the rod moves into the top of the radius link, above its centre position and takes the motion into reverse gear. Dependent upon the position of the piston in each cylinder, steam will be admitted to either the front or rear of the cylinder, fore gear moving the locomotive forward, reverse moving it backwards. Positions between the full forward, full reverse and the mid gear are shown on an indicator on the driver's reverser calibrated in percentage cut-off.

The piston valve spindle itself is connected, via the valve spindle cross head, to the radius rod, which in turn connects to the reverse shaft arm. This arrangement in itself would work, but would produce a gradual cut off profile, against the ideal which is for a sharp cut off and opening of the valve port. For this reason, two addition components are included, the vertically inclined combination lever, connected at its top end to the radius rod and the valve spindle

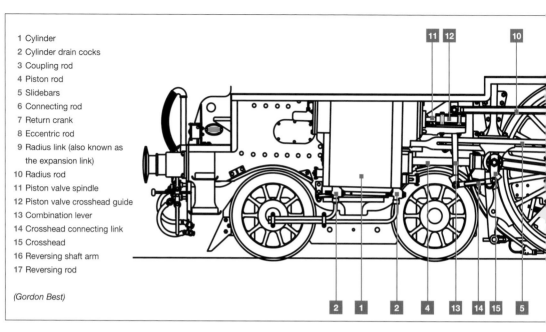

1 Cylinder
2 Cylinder drain cocks
3 Coupling rod
4 Piston rod
5 Slidebars
6 Connecting rod
7 Return crank
8 Eccentric rod
9 Radius link (also known as the expansion link)
10 Radius rod
11 Piston valve spindle
12 Piston valve crosshead guide
13 Combination lever
14 Crosshead connecting link
15 Crosshead
16 Reversing shaft arm
17 Reversing rod

(Gordon Best)

RIGHT A slide bar bracket casting before machining. *(David Elliott/A1SLT)*

BELOW A motion radius rod being forged at Heskeths of Bury. The forge hammer is in the background. *(Fastline Photographic/A1SLT)*

ABOVE The forged radius rods were then machined to the required dimensions at Ufone Precision Engineering. *(David Elliott/A1SLT)*

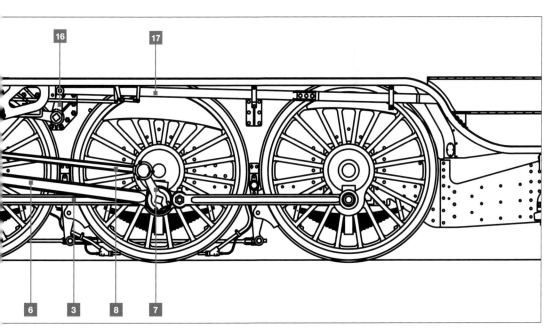

crosshead, and at its lower end via the union link to the main piston crosshead.

The motion components are a combination of forgings, castings and machinings from solid lumps of metal.

All the major rods were forged, i.e. they started from a billet of steel which was heated and forged to the required shape by drop hammers. This has the effect of 'stretching' the metal into the rod shape, which tends to align the crystals within the metal, giving it additional longitudinal strength. The material used for these rods was a medium carbon manganese steel, chosen to balance tensile strength with ductility. The ductility was an important factor, as in the event of a major failure the rods would tend to bend rather than snap, the latter having potentially much more dangerous consequences. Once forged, the rods were machined to dimensions. This was largely done using CNC (computer numerically controlled) milling machines which facilitate cutting the lightening flutes in the rods, the ends of each flute involving compound curvature.

The expansion links, being case hardened to provide hard wearing surfaces, had to have their curved slots ground to the finished size. The large size of these links necessitated the

LEFT Expansion link components after machining at Ufone Precision Engineering. *(David Elliott/A1SLT)*

BELOW The centre expansion link during precision drilling and reaming at Ian Howitt's Crofton Works. *(Nigel Facer/A1SLT)*

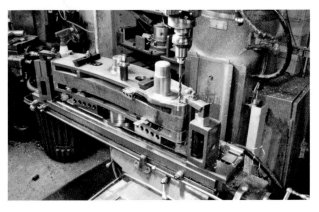

use of (what was then) the biggest CNC grinder in the UK, belonging to Bedestone Limited in Birmingham.

The solid parts – e.g. slide bars and cross heads were made from a medium carbon steel, chosen for its hardness and resistance to wear. Smaller motion components, and particularly any item which required case hardening, were made from a specially chosen case-hardening steel.

All motion components needed to be polished to reduce the risk of machining marks which could subsequently give rise to the initiation of fractures, but which also gives the characteristic smart appearance. This was done by hand, with the assistance of air-powered tools, mostly by the Trust's Peter Neesam, a key member of the construction team. This took some six weeks of dedicated work.

With the component parts manufactured,

attention could now turn to fitting – a critical part of the construction process. Despite the size of the components associated with the motion (including cylinders and axles), assembly of these items had to be within tolerances of thousandths of an inch, to ensure efficient operation.

The first essential was to ensure that the centre driving axle was set perfectly square to the frame and in line with the cylinder centrelines. This involved setting up a 'dummy' axle into the centre of the horns, and then taking careful measurements using an optical alignment system (provided and supported by the Severn Valley Railway), and using shims to make adjustments where needed. Once the position of the centre wheels was fixed, the other axle spacings then had to be set, with a centre-to-centre tolerance between each of the three coupled axles, of five thousandths of an inch. This was important, as with roller bearings, there is virtually no clearance in the axle bearings and any significant misalignment would result in premature wear and failure of the coupling rod bushes.

With these set up, the slidebars could be fitted and aligned. To do this a wire was stretched taut from the front centre of the cylinder, through the cylinder casing and back to the centre point of the centre driving axle. Measurements were made using a micrometer and an audible device, provided by Ian Howitt. This consisted of an electrical circuit which would trigger a buzzer when the circuit was made through the micrometer, allowing an accuracy considerably greater than relying

on seeing the device touch the wire. (The principle is identical to the game often seen at fairgrounds where a loop has to be moved along a twisted wire without touching the wire and triggering the buzzer.) Again, the position of the slidebars was adjusted using shims until perfect alignment was achieved.

The piston, piston crosshead, connecting rods and coupling rods could now be assembled. The next stage was to test this, which was done by means of jacking the locomotive up and lowering it on to a purpose-made 'rolling road' – a motor-driven set of wheels on to which the centre coupled wheels rested. The motor was switched on, transferring drive to the locomotive wheels. The hard work had all been worth it – the motion ran perfectly, the only noise being the rubbing of the crossheads on the slide bars as they bedded themselves in.

ABOVE Detailed
measurements were
taken for the eccentric
and radius rod lengths.
(David Elliott/A1SLT)

ABOVE RIGHT
Ian Howitt taking
measurements to
achieve the exact
centre line of the
cylinder for slide bar
alignment. *(David
Elliott/A1SLT)*

RIGHT The valve gear
motion was assembled
into position. *(Keith
Drury/A1SLT)*

The valve gear could now be assembled,
and the valve timing could be set up. Much
has been written on the subject of setting up
the valve timing, and many 'experts' have their
own views on how this should be done – some
using more complex methods than others.
Fortunately, the services of John Graham,

Chief Mechanical Engineer from NELPG (North
Eastern Locomotive Preservation Group) were
secured to set the valves.

As a starting point, with the reverser set in
mid gear and the piston at the centre of its
travel, the valves were set up so that both steam
inlet ports to the cylinders were closed. With
the valve chest end covers removed, the valve
spindle was marked with Tippex, which allowed
an easy mark to be made to identify the position
of the shaft. The gear was then slowly rotated,
and small adjustments were made to achieve
equal valve spindle movement in each direction.

Although seemingly a very simplistic approach
to quite a complex geometric system, the result
was highly successful, evidenced by the smooth
operation when the locomotive first moved under
its own steam, and the even beat which *Tornado*
produces both in forward and reverse motion.

RIGHT The complete
right-hand side motion
assembled. *(David
Elliott/A1SLT)*

Boiler

The boiler is the 'heart' of any steam locomotive, converting the heat content of the raw fuel (in this case coal) to high-pressure steam, which then passes to the cylinders to be converted into a mechanical motion. Although repairs have been carried out on numerous preserved locomotive boilers over the years, no locomotive boiler of anything approaching this scale had been built in the UK for fifty years.

It was decided that, in keeping with the principles of adopting modern manufacturing techniques, the boiler should be of an all-welded construction, a decision reinforced by the fact that the modern pressure vessel industry in the UK no longer builds riveted boilers.

Initial enquiries were sent out to around a dozen potential UK suppliers, but only two submitted a bid. Further discussions took place with both – but there were problems. One lost their design capacity due to the rationalisation of the industry at that time, while the other had a boiler designer with locomotive boiler design experience, but who unfortunately suffered health problems which forced retirement. The Trust could only put forward a 'general arrangement' based on the original riveted boiler design indicating overall dimensions and performance requirements – detailed design for the welded boiler would need to be carried out by the boiler manufacturer. It was considered essential for the design and construction to be by one company, to avoid the potential for split responsibility if anything subsequently went wrong.

Reluctantly, it was decided that the net would have to be cast wider, and enquiries were made in Europe. This identified three potential suppliers from former Eastern bloc countries, which was soon narrowed down to one – Dampflokwerk Meiningen in eastern Germany. The company, part of Deutsche Bahn, the German Federal Railway, had over 90 years' of experience, and was still building boilers of a similar size for heritage steam locomotives all over the world. Critically, they were able to demonstrate a comprehensive quality system with full traceability of all components, X ray facilities on all welds, and an experienced in-house design capability. This ticked off all the essential requirements for the Trust, but more importantly, for all the UK regulatory authorities.

With the selection of supplier made, the serious work of design and construction began. The Trust's requirement was to adhere to the original design of the A1 boiler as far as was practical within the confines of the change to an all-welded construction. With the supplier accepting responsibility for design and performance, they advised that they would not be prepared to fit a copper firebox, and this was accepted.

Essentially, the boiler consists of two major parts – the firebox, and the barrel. The firebox, as the name suggests, is where the coal is burnt to provide the heat. This is in two sections, the inner firebox (inside which the actual combustion takes place) and the outer firebox (which is the visible part when looking at the rear of the boiler from outside). The space between the two is part of the water space, where heat is transferred through the walls of the inner firebox to the water. The inner and outer sections have to be held in position relative to each other, and this is achieved by means of 'stays' welded at each end into the relevant box. These are only inches in length, and are subject to movement as the boxes expand at different rates due to temperature differences, and are therefore stressed components. For this reason, those on the upper part of the firebox sides and supporting the combustion chamber, which extends into the boiler barrel, use a ball and

BELOW **The front boiler barrel was rolled to shape through bending rolls at Dampflokwerk Meiningen in Germany, the chosen boiler supplier.** *(DB Werk Meiningen/A1SLT)*

RIGHT The outer firebox back plate during fabrication at Dampflokwerk Meiningen. *(DB Werk Meiningen/A1SLT)*

FAR RIGHT During one of the Trust's visits to Meiningen, Barry Wilson, Mark Allatt and Graeme Bunker in discussion while inspecting the firebox tube plate. *(DB Werk Meiningen/A1SLT)*

BELOW LEFT Some of the boiler stays await installation during the manufacture of the boiler. *(DB Werk Meiningen/A1SLT)*

BELOW RIGHT The firebox and foundation ring are seen being joined to the boiler barrel. *(DB Werk Meiningen/A1SLT)*

socket arrangement on the outer firebox, thus achieving a limited degree of movement. These are known as flexible stays. All the stays on *Tornado* were flush welded into the inner firebox, to allow for potential oil firing if this is ever required in the future.

Joining the inner and outer boxes at the bottom is the 'foundation ring' which rests on the main frames, thus supporting the back end of the boiler. Within the foundation ring sit the firebars, supporting the fire itself.

The firebox is welded to the barrel, the section next to the firebox being slightly tapered. This in turn is welded to the front cylindrical section. Within this barrel are a series of tubes, through which the hot gases from the fire pass on their way to the front smokebox. These are expanded into the front tube plate at the smokebox end of the barrel, and are welded into the inner firebox tube plate at the other end. The shell of the barrel is the steam generating space, partially filled with water, the

level of which is carefully maintained to ensure that the tubes and inner firebox are covered. The space above the water level contains steam. Two types of tubes are used, a small diameter tube to allow direct heat transfer through the tube wall to the water in the boiler, and a larger flue tube which, in addition to direct heat transfer into the boiler, contains superheater elements that add further heat to the steam after it leaves the boiler steam space on the way to the cylinders. *Tornado*'s boiler has a total of 121 small tubes and 43 superheater flues.

Towards the top centre of the boiler is a raised section – the dome – which houses the regulator valve and steam outlet pipe. The regulator valve controls the flow of steam out of the boiler and is the principal control governing movement of the locomotive (by allowing the flow of the steam to the cylinders). By taking steam off at the highest point in the boiler, the risk of damaging water carry over into the

steam pipe (known as priming) is minimised. *Tornado*'s boiler incorporates a further feature to reduce the risk of priming, introduced by Nigel Gresley on his Pacific locomotives during the 1930s, and officially referred to as a 'perforated steam collector' but commonly known as a 'Banjo dome'. This takes the form of an inverted tunnel section connected to and extending about four feet behind the dome. Under the tunnel are a large number of holes drilled through the boiler barrel. These allow steam to enter the dome from well clear of the regulator which reduces the tendency for water to be drawn into the regulator valve. Avoiding priming is very important, as any water carried over into the cylinders leads to the potential for serious damage.

Interestingly, the Germans had not come across this type of dome before, but later in the build, during one of the visits by members of the Trust, a document was noted referring to the 'Banjodom'. A new word had been introduced into the German language!

The inner firebox has a refractory arch (often referred to as the brick arch) fitted above the main fire grate, and extending backwards from the tube face of the inner firebox. This has several benefits – it partially protects the upper parts of the firebox from the direct and intense radiant heat from the fire, it slows the movement of the hot gases into the boiler tubes due to the need to pass around the arch, it allows time for complete combustion, and reduces the chance of continued burning of the gases once these are in the tubes, and it reduces the carry-over of partially burnt fuel as the particles hit the arch and are deflected back into the fire.

The arch on *Tornado* is made of refractory (high-temperature) cement, rather than bricks. This is cast in situ, by building a wooden former,

propped off the grate itself, and casting in the wet cement. The former is removed once the cement has fully set. The arch shape gives strength to prevent collapse, and this rests on lugs welded on either side of the inner firebox, which provide additional support. The arch typically lasts around 12 months in normal use.

Other fittings on the boiler include two safety valves: one designed to lift at the maximum boiler pressure of 250psi (pounds per square

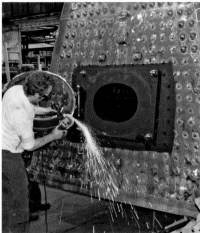

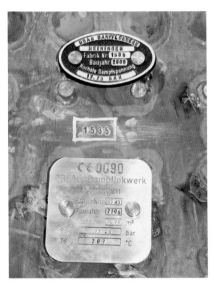

inch) and the other is set 1–2 psi higher, which prevent a potentially dangerous over-pressure situation. Also, sight glasses mounted in the cab on the back of the boiler give a visual indication to the crew of the water level in the boiler, mud hole doors (used during maintenance to allow access into the boiler shell for the removal of any accumulated sludge), and the water inlet pipes which admit fresh water to the boiler.

In the top of the inner firebox are three fusible plugs. These are bronze and are fitted with bronze pellets soldered in with lead. During normal operation these are cooled by the water in the boiler preventing the lead melting. If the water level in the boiler drops to a dangerous level, the plug is exposed to excessive heat and the lead melts, allowing the pellet to be blown out and alerting the crew to take urgent action on seeing or hearing the water/steam coming from the plug.

Manually operated blow-down valves are fitted to the front corners of the foundation ring. As steam is evaporated from the boiler any solids dissolved in the feed water are left in the boiler. Over a period of time these build up as a sludge, and if not subject to careful water treatment result in formation of scale on the firebox plates. This inhibits heat transfer, leading to overheating on the fire side of the plates and can cause potential damage to the boiler due to differential heat stresses. To prevent this, the valve is briefly opened periodically as part of the maintenance regime to flush out the solids. This helps keep the build-up of sludge under control and enables the boiler to be partially refilled with fresh water, which dilutes the dissolved solids, and enables the boiler to work longer between full washouts

Work on the boiler started following the signing of contracts in January 2005, and the completed unit was delivered to Darlington on 15 July 2006. During the construction period Trust representatives visited Meiningen on a number of occasions to review progress and discuss any issues arising from the build. An excellent rapport was developed with the build team during this period.

During the production of steam, heat is
added to the water to bring it to its boiling
point at the operating pressure of the boiler
(in *Tornado*'s case 250psi). This is 208ºC. At
this temperature, further heat is added to the
water (known as latent heat) to convert it to
steam. This steam (at the same temperature
as the boiling water) is known as saturated
steam. As the saturated steam is drawn away
from the boiler, and travels through the pipes
to the cylinders, some heat is inevitably lost
to the surrounding surfaces. The only way
for this to occur is for some of the steam to
give up its latent heat and turn back to water
(condense), which results in a steam/water mix
at the cylinders (known as wet steam). This is
undesirable for a number of reasons, and limits
the efficiency of the locomotive. The problem is
overcome by adding more heat to the saturated
steam, which is known as 'superheating'. This
allows some of the superheat to be lost, without
the condensation back to water. Also, one of
the major sources of inefficiency in a steam
locomotive is the amount of latent heat that is
lost up the chimney with the exhaust steam. A
further benefit of superheating (typically, up to
380–420ºC), is that less steam is needed per
horsepower produced, and thus less steam is
used and therefore less latent heat is wasted.

To achieve superheating the steam drawn
off through the outlet steam pipe goes into a

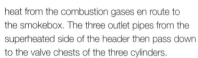

RIGHT **The superheater header casting is machined.** *(David Elliott/A1SLT)*

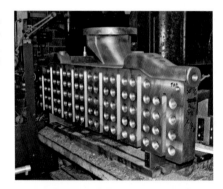

RIGHT **With the header fitted to the front of the boiler, the superheater elements were placed in position in the boiler flue tubes.** *(David Elliott/A1SLT)*

BELOW **As delivered – the castings for the firegrate.** *(David Elliott/ A1SLT)*

BELOW RIGHT **The part-assembled firegrate, showing the rocker mechanism operating rods.** *(David Elliott/A1SLT)*

superheater header at the front of the boiler. This is a complex casting which, as well as the connection for the steam admission pipe, has the flow and return connections for the superheater pipes (known as elements), and the final outlet connections for the superheated steam to the cylinders. Each superheater element is folded back on itself within the flue tube so the steam makes four passes through the flue, gaining

heat from the combustion gases en route to the smokebox. The three outlet pipes from the superheated side of the header then pass down to the valve chests of the three cylinders.

Fire grate and ashpan

The fire grate consists of a series of low-grade cast iron bars, positioned parallel to each other, and spaced so the coal is retained on the grate during combustion, but the smaller ash particles fall through to the ashpan below. The grate on *Tornado* is of the 'rocking type', with the individual bars linked together by connecting rods, which are in turn linked to levers in the cab. The grate is split into left and right-hand halves, each with its own operating lever. The levers are fitted with locks to prevent inadvertent operation while detents enable small movements of the bars to be made to assist in keeping the fire clean. On completion of a run, the detents can be disengaged to enable the bars to be rocked to the full extent, allowing the fire to be dumped in its entirety. Most of the fire bars were cast from patterns originally made for preserved A2 class No. 60532 *Blue Peter*, which actually carries an A1 boiler, and has a similar grate system.

The ashpan is a fabricated steel box, which fits underneath the fire grate and between the rear part of the main frames. The box is shaped, with sloping sides, to allow easy removal of the accumulated ash within the box. Two cast steel doors are fitted to the ashpan, operated by a

lever at track level on the fireman's side of the rear frames. This enables the accumulated ash to be dropped into a pit with minimal assistance from fire irons. The front of the ashpan is fitted with a hinged door which forms the damper. The intensity of the combustion on the grate is controlled by the amount of opening of the air damper.

Combustion of fuel uses air from two sources. Primary air comes up through the fire grate and enables the coal to burn and liberate carbon monoxide and hydrogen which is then burnt in the form of visible flame in the firebox. In order to ensure as near complete combustion as possible, further air (known as secondary air) is admitted via the flap in the firehole door.

The damper regulating the primary air has three open positions in addition to fully closed, controlled by a lifting handle to the right of the firehole door. Even when fully closed, gaps at either side of the damper door provide enough air to keep the fire alight when the locomotive is idle. When the air damper is open, and the locomotive is travelling under power, substantial air is drawn through the bed, producing a very hot (white) fire, with intense heat release and almost fully complete combustion.

The fire door flap is used to control the colour of the exhaust which should be a light grey haze. Too little secondary air causes black smoke due to incomplete combustion – too much secondary air causes the combustion gases to be cooled excessively and runs the risk of damage to the firebox due to localised cooling. The ashpan is fitted with manually operated water quench

sprays, to reduce risk of buckling and damage to the ashpan plates due to heat.

Emptying the ashpan following a run is normally carried out using a hosepipe to damp down the ash to prevent it blowing about and contaminating the motion bearings and the paintwork. It is also necessary to use a hose or rake in the ashpan to remove all the ash from areas where the shape of the ashpan prevents it from falling out by gravity.

ABOVE Part of the assembled grate fitted inside the firebox. *(Barry Wilson/A1SLT)*

LEFT The ashpan during fabrication. *(David Elliott/A1SLT)*

BELOW LEFT The inverted ashpan, showing the ash dump/ primary air inlet chutes. *(David Elliott/A1SLT)*

BELOW The completed ashpan at Hopetown Works, ready for fitting to the boiler. *(David Elliott/ A1SLT)*

RIGHT The smokebox door hinges were bent into position and welded at Ian Howitt's Crofton Works.
(David Elliott/A1SLT)

RIGHT The completed smokebox door ready to leave Crofton Works.
(David Elliott/A1SLT)

RIGHT The door as fitted to the smokebox barrel and awaiting further progress at Hopetown Works.
(David Elliott/A1SLT)

Smokebox, smoke deflectors and exhaust

The smokebox, joined to the front of the boiler barrel, is a rolled steel section, with a welded seam along the top to form the circular shape. Its overall diameter is slightly larger than the boiler barrel itself, which allows for the provision of the insulation and cladding to the boiler, whilst still aligning with the smokebox barrel. The smokebox is bolted to the boiler, using round-headed bolts to simulate the rivets which were used on the original Peppercorn A1s.

In order to accommodate the difference in diameters between the two sections a spacer ring is inserted around the join – this was made in three sections to facilitate fitting. The boiler barrel was supplied with the fixing holes pre drilled, but the holes had to be continued through the spacer ring and the smokebox barrel, and in order to ensure perfect alignment for the close fitting bolts, the drilling had to be carried out with the sections in place. Using a heavy duty drill in the confined space was not going to be the easiest of tasks, so a special jig was made to support the drill as it was moved around the circumference to drill each of the holes in turn.

On the front of the smokebox is the hinged smokebox door, designed to allow access for maintenance and cleaning. The door itself was manufactured by a specialist metal-forming company, using a combination of 'spinning' and hand-forming. A blank disc of metal was spun into shape using a pair of rolls, one on either side on the plate, guided by a CNC system, to produce the basic dome profile. The edge curvature was then created by heating the metal locally with an oxy-propane torch and gradually hand-forming it over a die, made to the required shape. Finally, the shaped unit was machined on a vertical borer to give the correctly finished sealing face.

The door is secured by two hinges on to a mounting ring, which in turn is welded (riveted on the original A1s) to the front end of the smokebox barrel. The mounting ring itself is 'U'-shaped to fit into the barrel, starting out as a single ring of metal, into which the groove was machined to correspond to the profile of the original design. A further grooved ring was machined to profile and welded to the front of the smokebox ring to accommodate a glassfibre rope to provide a tight seal and prevent air ingress when the door is closed. The original A1s would have used asbestos rope, but obviously this is no longer acceptable for safety reasons. The door securing mechanism is a large threaded 'T'-shaped bolt (known as the dart), which passes through an opening in a crossbar located on lugs across the front of the smokebox barrel, and turns through 90°, such that it is held by the bar. It is tightened on the outside by means of the two levers, resembling the hands of a clock face and characteristic of the front of all steam locomotives. The shorter (rear) lever rotates the dart such that it is in the engaged position with the lever in the 6 o'clock position. The outer lever is attached to a nut on the front end of the dart and is used to tighten the smokebox door on to the seal.

The primary function of the smokebox is to collect the waste gases from the boiler, and to eject these via the chimney. In order to bring this about the exhaust steam from the cylinders is ejected via a nozzle (the blast pipe) in the base of the smokebox. Above the blast pipe is the chimney cowl, which forms part of the chimney, and extends downwards into the smokebox. The force of the steam being ejected through the nozzle draws the surrounding waste gases into the cowl, and they are then carried upwards and ejected through the chimney. Many locomotives use simple single or double blast pipe arrangements, however this does have limitations, particularly on larger locomotives travelling at high speed, where the inability to get the gases away efficiently will impact upon boiler performance.

To overcome this, *Tornado*, in common with the original Peppercorn A1s, and many other classes of express locomotive, uses the Kylchap exhaust arrangement. This was developed and patented by French steam Engineer André Chapelon, and uses a second-stage nozzle designed by Finnish Engineer Kyösti Kylälä, known as the Kylälä spreader. Hence the derivation of the name 'Kylchap'.

The general principle behind the system is that instead of using the single blast pipe (or double in the case of a double-chimney locomotive), with the single entry point for the waste gases, the Kylchap system uses a three-stage system with the waste gases entering at three points in the vertically stacked arrangement. This gives a more even flow of gases into the exhaust, thus reducing back pressure and improving the overall efficiency of the boiler.

Each of the two blast nozzles bringing exhaust steam from the cylinders is fitted with dividing lugs at the exit of the pipes, partially splitting the steam into four jets, which enter the second stage, entraining some of the exhaust gases in the process. The second stage, the Kylälä spreader, consists of four slightly convergent tapered tubes, which combine into a single parallel cowl, the bottom of which is flared like a trumpet. This forms the second point of entrainment. The single cowl then exhausts into a the larger chimney cowl which is initially a convergent taper and then changes to a divergent taper up to the chimney top. The third point of gas entrainment is on the entry to the chimney

LEFT Cast exhaust system components ready for further working.
(David Elliott/A1SLT)

LEFT The exhaust second-stage Kylälä spreader demonstrating the 'four-tube' arrangement.
(David Elliott/A1SLT)

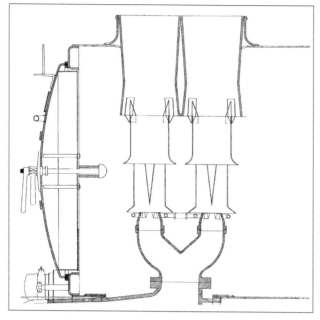

ABOVE Schematic arrangement of the Kylchap exhaust system. *(David Elliott)*

with a lip and one without. These were made as castings, machined to final dimensions. Most of the original Peppercorn A1s started with unlipped chimneys, possibly as an economy measure due to the availability of materials in the post-war period. However, from 1950 these were gradually changed to the lipped type. The exact reason for the change is a little uncertain – it may have been partly for aesthetic reasons, but it is more likely that it was to improve the smoke dispersal from the chimney – the lip acting to break up eddy currents which tended to form around the unlipped versions, causing smoke to down draft around the smokebox area.

In order to ensure the efficient exhaust of gases when the locomotive is stationary (i.e. no exhaust from the cylinders), the top end of the primary blast pipes have a circular pipe around the outside – drilled with a series of equidistant holes. These are connected to the live steam supply via a control valve in the cab. This is known as the blower and, when turned on, the live steam passing through the jets produces the same induction to draw the waste gases through the boiler as is normally experienced with the exhaust gases from the cylinders. The driver also uses the blower on occasions to prevent fire blowback.

The smoke deflectors are fabricated plates,

cowl. This arrangement is shown in the drawing above.

The double chimney itself had to be redesigned in order to stay within the specified 13ft 0in height limit (the original Peppercorn A1s were 13ft 1in). Two chimneys were made, one

RIGHT A lipped chimney casting at Taylors, South Shields. *(David Elliott/A1SLT)*

BELOW The smokebox, showing the horizontal bar arrangement which is used to secure the smokebox door. *(Rob Morland/A1SLT)*

RIGHT The exhaust blastpipe showing the positioning of the blower rings, used to induce boiler draft when the locomotive is stationary. *(David Elliott/A1SLT)*

secured to the outside of the smokebox with brackets. Whilst also enhancing the appearance of the locomotive, these were primarily installed for practical reasons. On many express locomotives travelling at speed, the smoke from the chimney tended to be drawn down around the cab area, reducing forward visibility for the crew, and presenting a safety hazard when watching for signals. The smoke deflectors, particularly at speed, cause an air movement around the sides of the locomotive which encouraged the smoke to travel directly backwards from the chimney, above the boiler and cab, giving much improved visibility. On *Tornado* (again in common with the original Peppercorn A1s), they also conceal other smaller components such as the electrical generator.

Most steam locomotives had a manually cleaned smokebox – after a certain number of hours of operation (on long-distance trains usually after every journey), the smokebox door would be opened and the accumulated ash (carried through from the fire, but not ejected through the chimney), would be shovelled out of the smokebox. This was a time-consuming and arduous task. In addition, running steam locomotives now requires the fitting of a spark arrestor to prevent particles of burning coal being emitted from the chimney, and potentially causing lineside fires.

To deal with these issues it was decided to fit *Tornado* with a 'self-cleaning' smokebox, combined with a spark arrestor. This system had been tried successfully, and was a standard

inclusion on a number of later BR locomotive builds, and the design concept for the system used on *Tornado* replicated that used on preserved BR Standard 'Britannia' class locomotive No. 70013 *Oliver Cromwell*. A metal plate inclined forwards from top to bottom and extending approximately three quarters of the way down into the smokebox, is fastened in front of the superheater header, such that the waste gases leaving the boiler, hit the plate and are deflected downwards. Any particulate matter hitting the plate is broken down into smaller particles. The positioning of the plate is important, to minimise any back pressure due to the extra resistance, which would impact on boiler performance.

The plate then has a horizontal section which forces the gases towards the front of the smokebox around the blast pipe at low level with further impacts continuing to break down any larger particles. Finally, a mesh screen is fitted sloping from bottom to top towards the front of the smokebox, and linked to the lower horizontal plate. This forces the gases to pass through the mesh screen to the exhaust arrangement. Any oversize particles are stopped by the screen, and fall back down to be further broken down as they are swirled around the metal sections. This system has three major benefits: first, any particulate matter is broken down into very small pieces, which, even if still at combustion temperature when they leave the chimney, are rapidly cooled in the air, and are certainly below combustion temperature when they hit the ground. Secondly, the longer residence time in the smokebox acts to give the particles longer to cool down before being exhausted through the chimney. Finally, the breaking down of the larger particles means that these are light enough to be carried away in the exhaust gases and very little residue is left in the smokebox – hence the 'self-cleaning'. Experience has shown *Tornado* leaves typically, one dustpan full of ash in the smokebox after a long run!

Front bogie

The role of the front bogie is to assist in guiding the locomotive on curves and other track irregularities such as points and crossovers. However, it also plays an important role in carrying the overall weight. On *Tornado* the front bogie carries a load of approximately 19 tons (over the two axles), the main driving axles carry loads of between 22 and 23 tons each, and the trailing axle (Cartazzi), carries a load of around 19 tons, maintaining the locomotive within the maximum axle loading permitted on the present railway network of 25 tons.

The bogie is constructed such that it can swivel around a centre pin, but can also move from side to side, against strong springs which provide a restorative force attempting to keep

the bogie in the centre position – this assists in guiding the locomotive on curves.

The basic bogie construction is similar in concept to that of the main frames. There are two main frame plates, separated by three cast steel frame stays. The front and rear stretchers are known as the bogie frame stays. The axle box horn guides are integral parts of these frame stays. The axles are again contained in cannon boxes, as described for the driving wheelsets, which in turn accommodate the Timken roller bearings. These are then fitted into the horn guides, secured on the underside with bolted horn stays to prevent the assembly dropping out if the bogie is lifted. Wheel and axle construction is identical to that of the main driving axles, the wheels in this case being 3ft 2in in diameter.

The third frame stay, known as the bogie bottom centre, is a large casting which provides the major part of the structural integrity of the bogie. Within the bogie bottom centre is the bogie crosshead which houses a bearing bush which engages with the 'pintle' attached to bogie top centre and hence to the locomotive frames. This allows the bogie to rotate. The crosshead also slides sideways and is restrained by the side control springs which are housed in pockets in the bogie bottom centre.

When the bogie is positioned dead centre, each of the springs has a compressed force of 4 tons, acting in opposing directions, and therefore maintaining the centre position. As the bogie starts to move to the side, the springs on that side are compressed. This builds, at maximum compression, to a restorative force of 7.5 tons, trying to push the bogie back to the central position.

The outer extremities of the bogie top centre casting are fitted with de Glehn-type side bearers which transfer the weight to the locomotive directly from the main frame plates into the bogie frames. These comprise two flat steel plates on the top of the bogie bottom centre. On each plate a bronze block with a flat bottom and a spherical depression on the upper side is free to slide laterally. A convex spherical pad is bolted on to the bogie top centre which rests on the bronze block. The spherical components allow the bogie to see-saw longitudinally to accommodate track imperfections.

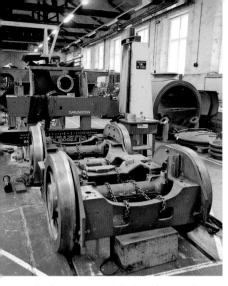

Inadequate strength in the side control springs in the original Peppercorn A1s was one of the main reasons that a number of these locomotives gained a reputation for rough riding. Over their years in service the springs were gradually increased in strength up to the figures quoted above, which has largely eliminated the problem.

Cartazzi axle

The locomotive rear axle uses axle boxes of the Cartazzi design. The principle, invented by F.I. Cortazzi, (subsequently known as Cartazzi) who worked under Archibald Sturrock on the Great Northern Railway in the mid-19th century, uses axle boxes and horn guides which, when viewed in plan, are angled forward from the normal rectangular position. In addition, the upper surface of each axlebox has a slipper block with two opposed inclined planes arranged such that with corresponding inclines on the spring plank, the rear of the locomotive is lifted slightly as the wheelset moves away from the centre position. The weight acting on the incline provides a centring force which attempts to restore the wheelset to the straight-ahead position. In this position, the spring plank is resting on both of the opposed inclines resulting in no side force. Numerous refinements have been made to the principle since it was invented by Cartazzi, not least to allow its use with roller bearings.

The major advantage of using this system on *Tornado* (and the LNER Pacifics in general)

was that it takes up much less space under the ashpan than the alternative, which is a pony truck. The latter works on a similar principle to the front bogie, moving around a centre pivot. With a large firegrate, and consequently a large ashpan, this was a very important consideration.

The axlebox horns are fitted to the outer main frames, and the axle itself passes through a clearance slot cut in the inner main frames.

Springs

Springs are provided between all the wheel axles and the frames on both the locomotive and tender. This allows the axles a small amount of vertical movement relative to the frames, which improves the ride and reduces track wear, but primarily improves safety by reducing the risk of derailment.

In the case of the front bogie, these are coil springs, two at each end of each axle. The springs rest on brackets that form part of the bogie bottom centre and the bogie frame stays. The weight on the bogie is transferred to the springs via spring planks which are located on the outer ends of the cannon boxes, through spring bolts which pass through the centre of each spring, and have spring cups which locate the springs and contain secondary rubber springs.

The rest of the springs, for the main coupled wheelsets, the Cartazzi wheels and the tender wheels, are all leaf springs. These consist of multiple leaves of spring steel (16 per spring in the case of the main coupled wheels, 14 in the case of the Cartazzi wheels, and 11 for the tender wheels). The leaves vary in length with the longest at the top of the stack, and gradually reducing to the bottom.

The main coupled wheel springs are below the axles. The ends of the springs are located into cast spring hanger brackets which are riveted on the inside of, and hang below, the

TOP Cartazzi axle springsets showing the leaf and clamping arrangements. *(David Elliott/*

ABOVE A tender spring in position indicating the hanger arrangement. *(David Elliott/A1SLT)*

RIGHT A close-up of the completed tender showing the spring and brake block arrangements. *(David Elliott/A1SLT)*

main frames. The spring brackets carry spring cups containing secondary rubber springs through which the spring bolts pass. The upper end of each spring bolt locates on a specially shaped washer which sits on the outer ends of the springs. The spring leaves are held together by a buckle with a long rivet through the middle of the leaves. The buckle has a clevis on the upper side which is connected to the underside of the axle/cannon box by way of a spring link and pins.

The Cartazzi and tender springs are mounted above the axles. The spring outer ends have a lug which fits into a corresponding shape in the spring hanger castings. These are in turn bolted to brackets which are attached to the main frames. A lug on the top of the axle boxes acts directly on to a lug on the base of the clamp around the centre of the spring assembly. Thus the axles are trying to push the spring upwards, whilst the weight of the locomotive and tender are pushing downwards though the spring hangers.

Brakes

The original Peppercorn A1s used a steam braking system on the locomotive and tender, with vacuum braking for the train, the vacuum being provided by a Davies & Metcalfe steam ejector and gradable vacuum-operated steam brake valve.

All modern railway vehicles are air braked, but with a number of heritage coach rakes still being vacuum braked (principally on heritage railways, but also a limited number on the main line), a decision was taken that *Tornado* would have a dual braking system. The primary system would be air based, but a slaved vacuum system would be provided for the occasions when the locomotive hauls vacuum-braked stock. This introduced the requirement for additional braking controls, which as far as possible needed to be automatic in nature, but also the call for additional pipe work as both air and vacuum pipes had to be provided to the rear of the tender for coupling up to the train.

On the air braking system, the locomotive and tender are treated as separate vehicles in terms of braking, each having its own air reservoir and distributor, which are located under the cab and under the back of the tender respectively.

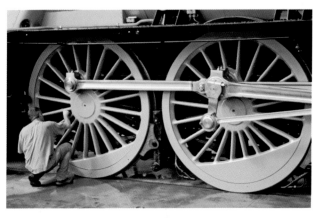

Compressed air is provided by two steam-driven, in-line, compound compressors each having one steam cylinder and two air cylinders, giving two-stage air compression. The compressors were bought from DB Meiningen (the boiler manufacturer) and are of East German origin, but based on a Westinghouse design. Both are located between the locomotive frames, one immediately in front of the firebox, while the other is mounted slightly further forward on the star stay, which required substantial modification to allow the compressor and air brake cylinders to be fitted.

After passing through pipework forming a cooling radiator and various water and oil separation vessels, the compressed air at a pressure of approximately 140psig is stored

ABOVE Taken during the painting of the wheels at Hopetown Works, this photograph shows the coupled wheel spring and brake block arrangements. *(David Tillotson/A1SLT)*

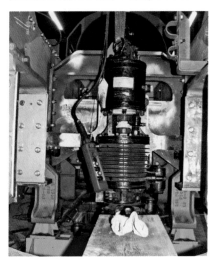

LEFT The front air brake pump being mounted between the locomotive frames. *(David Elliott/A1SLT)*

in four main reservoir tanks arranged in series and located under the tender tank. The main reservoir feeds each of the two brake reservoirs via non-return (check) valves.

On an air-braked system, each vehicle in the train has a separate reservoir. With the BR two-pipe air-brake system the pressure in each vehicle reservoir is maintained by the 'reservoir pipe' fed from the main reservoir via a 140psig to 100psig pressure reducing valve on the locomotive. The hose couplings and cocks on the reservoir pipe are painted yellow for easy identification purposes. Brake control is by the brake pipe which passes the full length of the train and can be identified on vehicle ends by red-painted hose couplings and cocks.

The brake pipe under running conditions, is maintained at an air pressure of 72.5psig by the driver's automatic air-brake valve. This pressure maintains the brakes in the fully released condition.

The main control in the locomotive cab is a Westinghouse M8A automatic air brake valve. When the brake handle is operated under normal circumstances it is self-lapping, which means that the pressure in the brake pipe is directly proportional to the position of the handle. As the handle is moved from normal (released) position through 'initial' to 'full service' application, allowing the air pressure in the brake pipe to reduce, the brakes are applied progressively, being full on when the air pressure in the brake pipe drops to approximately 46–48psig. Simultaneously, the distributors on the locomotive, tender and train vehicles react to the drop in brake pipe pressure by proportionally feeding air from the brake reservoirs to the brake cylinders, which push the brake blocks on to the wheels. The pressure in the brake cylinders for a full application is 45psig for the tender and 55psig for the locomotive, controlled by the distributors which are set up individually for the design cylinder pressure on each vehicle.

To prevent sudden jerks or damage to couplings, the rate of fall of brake pipe pressure is carefully controlled by the driver's brake valve. For a more rapid brake application, the brake handle can be moved further to an emergency braking position, where the brake pipe is opened fully and quickly to atmosphere, leading to a rapid drop in air pressure and an immediate full application of the brakes. This situation would also occur if the train pipe fractured or became disconnected for any reason, thus the system is effectively 'fail-safe'. The brakes are released by rebuilding the air pressure in the brake pipe which causes the distributors to release the air in the brake cylinders to atmosphere.

The air pressure in the train pipe is monitored both from the driver's cab and the guard's compartment in the train. When a locomotive is connected up to a train, a mandatory brake test must be carried out to establish the integrity of the braking system prior to movement. The process is for the driver to release the brakes while the guard monitors their gauge. The guard then makes a brake application using the emergency brake valve in the guard's compartment. The driver observes that the locomotive brake pipe pressure drops and the locomotive brakes are applied. This proves the continuity of the brake pipe which is essential to ensure that the entire train has the automatic brake operative.

When hauling vacuum-braked stock, the locomotive and tender continue to be air braked, but the train vehicle brakes are now operated by vacuum – this is again a fail-safe system in that vacuum is applied to release the brakes, and if the vacuum fails the brakes are applied.

The vacuum is created by a twin-barrel steam ejector mounted on the driver's side of the smokebox, behind the smoke deflector. This operates from boiler steam. The small ejector is used to maintain the vacuum against leakage while the large ejector is used as required to assist in rapidly releasing the brakes. The normal vacuum when the brakes are released is 21in of mercury (approximately 10psig below atmospheric pressure). When the vacuum braking system is initiated, air/vacuum isolating valves detect the presence of vacuum and cause the brake distributors on the locomotive and tender to react to changes in vacuum instead of air. In order to revert to air control, it is necessary to dump the vacuum in the small distributor control reservoirs by operating the 'dagger' valves in the cab and on the front of the tender. The reason for having the locomotive air brakes controlled by vacuum is to ensure that the locomotive brakes are

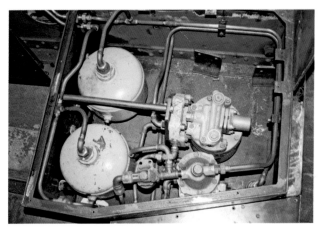

ABOVE Part of the brake control gear located in a cubicle under the cab floor. *(David Elliott/A1SLT)*

applied in event of a brake application from the train, be it due to a break in the brake pipe, an emergency application by the guard, or use of the communication cord.

The vacuum is controlled by an air/vacuum proportional valve which detects changes in the air brake pipe caused by use of the driver's air brake valve and causes the vacuum to reduce proportionally to the pressure in the air brake pipe.

A further brake control is activated by the TPWS/AWS systems (see electrical systems section for a further description of these). This consists of a Baldwin electro-pneumatic valve which under normal running condition is maintained in the closed position by the TPWS/AWS systems. If these are triggered, then the valve opens, dumping air pressure in the air brake pipe and thus providing a full brake application

There is also a 'straight air' brake which is used for shunting and coupling up to trains. This is in the form of a black-handled direct-operating brake valve which controls air directly to the locomotive and tender brake cylinders. Because it is direct acting the straight air brake is very quick acting and sensitive, however, it must not be used in normal operations as it is not automatic, and in the event of a broken air pipe or a fault in the brake valve, the braking effect will be lost altogether.

The air brake cylinders themselves, which apply and release the brakes, operate through a series of mechanical rods and joints to operate the brake blocks on to the wheel tyres (on the locomotive this is on the main coupled wheels

only, not on the bogie or Cartazzi wheels). The brake blocks are made of a high-phosphorous cast iron, designed to be softer than the metal of the tyres, whilst at the same time giving maximum resistance to wear. Nevertheless, these items are subject to fairly frequent change – typically three times a year on the locomotive and once per year on the tender.

A handbrake operates on the tender brakes only, and is located on the cab end of the tender. This is a mechanical device which operates directly on to the cross shaft of the tender brake linkage, applying the brake blocks to the wheels.

Injectors

The injectors are the primary means of transferring feed water from the tender to the boiler.

As these operate on steam, clearly they cannot work when there is no steam pressure, so one of the blowdown valves on the boiler is fitted with a fire hose connection, allowing it to be filled from cold from a fire hose.

Once steam is available (above about 50psig) then the injectors are able to operate. *Tornado*, in common with the original Peppercorn A1s (and most steam locomotives), is fitted with two injectors, one operating on live steam only, and one operating on either live steam or exhaust steam from the cylinders. The exhaust steam injector makes considerable savings by using the residual energy in the steam leaving the cylinders (up to about 7 per cent saving in water and fuel, compared with using live steam only).

The injector works by passing the steam though cones (typically three), during the process of which it induces water, combines with it, and achieves a force at which it can overcome boiler pressure to enter the boiler (towards the front end) though a non-return valve, known as the clack valve.

For the live steam injector, the Trust was lucky to acquire two new Davies & Metcalfe Monitor injectors, from a batch originally built for the Southern Railway's Bulleid Pacifics that had been rescued from Barry scrapyard devoid of most of their non-ferrous fittings. Although these had 12mm cones, as opposed to the 11mm on the original Peppercorn A1s, this was

considered a bonus, and has proved to be so, giving excellent water flow and reliability. The main problem was that these were vertically orientated units, as opposed to the normal horizontal units on the original Peppercorn A1s, and this did require quite a bit of reconfiguration of pipework to fit. This injector is located on the driver's side of the locomotive.

The normal method of starting an injector is to manually open the water valve on the front of the tender (the water initially flowing through the overflow), and then gradually open the steam valve until it is heard to be picking up the water (evidenced by a roaring sound), and then if necessary, close the water valve down until just a trickle is still leaving the overflow. The injector on *Tornado* is so efficient that there is no need to further adjust the water valve – the steam flow picks up almost all the water available.

The exhaust steam injector, fitted on the fireman's side of the locomotive, was a Davies & Metcalfe Type J on the original Peppercorn A1s. However, the Trust was able to acquire a specimen of the earlier Type H injector from A4 class No. 60007 *Sir Nigel Gresley* whose owners had decided to fit two live steam injectors to better suit that engine for use on heritage railways. Both types were fitted with 10mm delivery cones. The main difference between the two types is that the Type H injector has an automatic water valve while the later Type J has a manual valve. In practice, with the Type H once the tender water valve is set up to eliminate overflow, it can be left at that setting all day.

The injector is started on live steam, once it detects exhaust steam is available (by monitoring the pressure from one of the main steam pipes to the cylinders) a shuttle valve operates to use primarily exhaust steam (around 90 per cent) with a small top-up (around 10 per cent) of live steam.

Pipework

Whilst less glamorous than the larger components, *Tornado* is fitted with a mass of pipework. This is needed to get the water into the tender, to transfer the water to the boiler (via the injectors), to take the steam from the boiler to the cylinders (and to its various auxiliary users), to distribute the compressed air and vacuum for the braking systems, to distribute lubricants to the points required, and to carry cables to all the electrical equipment fitted to the locomotive.

Although some of this was in common with the original Peppercorn A1s, the additional features on *Tornado* make this much more extensive – the braking system pipework is doubled due to the need for separate air and vacuum braking systems, and the electrical conduits are almost all additional – the electrical equipment on the original Peppercorn A1s being very limited. In order to preserve the appearance of the locomotive as near as possible to the original Peppercorn A1s, it was necessary to hide as much of this additional pipework as possible, and this caused a number of problems due to the very limited space available within the frames on both the locomotive and tender, due to other major components.

In general, most of the pipework on the locomotive itself is in a heavy grade copper,

with many of the joints using bespoke fittings or heavy duty Yorkshire fittings (similar to domestic-type Yorkshire fittings but using silver solder as opposed to soft solder). Steam heating and vacuum brake pipework is mostly in mild steel, with screwed or welded joints as appropriate. The connections between the locomotive and tender are in flexible piping, and in the case of electrical connections are fitted with weather-proof plug and socket arrangements to allow the tender to be separated when required.

ABOVE Exhaust steam injector pipework being fitted through the inside of the frames. *(David Elliott/A1SLT)*

LEFT The vacuum brake train pipe threading its way through the frames. *(David Elliott/A1SLT)*

FAR LEFT This demonstrates the complexity of the pipework fitting through the frames. *(David Elliott/A1SLT)*

LEFT Exhaust steam injector pipework showing complex shape necessary to allow fitting through the frames. *(David Elliott/A1SLT)*

Inside the boiler the pipework is steel with welded fittings. From the regulator in the dome, the main steam pipe runs forward to serve the cylinders via the superheater header. Further smaller steam pipework feeds a small manifold inside the boiler from which additional pipes feed the two injector valves on the boiler backhead, a feed out of the front tubeplate which serves the lubrication atomisers, and the chime whistle stand. Finally, one pipe goes to the external manifold (the steam stand) located high on the rear of the boiler in the cab. This has a series of valve-controlled off-takes to serve the various auxiliary steam users around the locomotive.

Facing the backhead, the first of these (left-hand) supplies steam to warm the mechanical lubricator to keep the oil viscosity low and allow it to flow efficiently. The second serves the steam vacuum ejector for the vacuum brake system, and the third serves the exhaust blower ring. The outlet pipe from this valve travels straight through the boiler barrel itself. A smaller supply off this pipe serves the standard LNER whistle.

The next valve supplies the rear air pump, followed by the turbo alternator supply. The next off-take supplies the steam pressure gauge in the cab, with the final valve supplying steam to the front air pump.

There is also a further steam valve attached to the exhaust (right-hand) injector control valve which supplies the carriage warming

steam. Originally, this valve supplied the turbo alternator and the valve presently used for the turbo alternator was used for the steam heating, However, the uprated turbo alternator tended to be throttled by its valve, whilst the larger steam heat valve was difficult to regulate when only one vehicle (the support coach) is being steam heated. To improve matters the pipework has been swapped over. (The layout of the steam pipework and control valves in the cab are shown in more detail in the later controls section.)

The steam heating control system is very basic, relying on the fireman to regulate the steam pressure using the steam heat valve. If the pressure rises above 65psig a safety valve at cab floor level opens to alert the fireman to the excessive pressure. Steam for heating is usually regulated to between 30 and 50psig.

Part of the regulatory requirements for high-speed running on the main line are that the locomotive should have an audible two-tone warning capable of being heard at a significant distance. There was initially some concern that the original A1 whistle would not be considered adequate for this requirement, and the Trust gratefully accepted the long-term loan by a covenantor of an A4 chime whistle, which started life on No. 60023 *Golden Eagle*. This is now a familiar sound when *Tornado* is on its travels. The Trust is in the process of having its own chime whistle made.

Cab

In order to give recognisable indications of progress to the covenantors, the basic cab structure was put together in the early days of construction, and features in many of the photographs from that era.

However, as design progressed, and in particular as the team set about the task of concealing the extensive amount of additional equipment which would be an essential feature of *Tornado* (air-brake control equipment and electrical equipment being the two prime examples), substantial modifications to the original structure were necessary.

On the original Peppercorn A1s the cab floor was simply made of angle iron, bracketed to the frames, infilled with steel plating and overlaid with wooden planks. On *Tornado* it

RIGHT The inverted cab structure at Hopetown Works, prior to riveting. *(Barry Wilson/A1SLT)*

FAR LEFT TOP
Inverted cab showing the edge detail down the side. *(David Elliott/ A1SLT)*

FAR LEFT BOTTOM
Cab being riveted in the workshops of the North York Moors Railway at Grosmont. *(David Elliott/A1SLT)*

LEFT The riveted cab at Grosmont prior to return to Darlington. *(David Elliott/A1SLT)*

was decided that the air-brake control gear could be located in a specially constructed and shaped cubicle which would form the basic floor structure, and which would be overlaid with hardwood floor panels. The upper part of the cab structure (side panels and roof) are constructed from steel plate cut to shape and reinforced by steel angle sections. Jointing is a combination of welding and riveting, to give the appearance as per the originals. The upper structure and floor are joined to create a single unit, and the extra rigidity offered by the box unit floor allowed design changes to be made to allow the removal of the cab unit in one piece – a major benefit during subsequent maintenance. The complete cab unit can be removed in less than five hours.

All pipe connections (11 in total) passing into or through the cab unit are fitted with unions to allow easy disconnection, and all electrical cables are fitted with military-specification multi-pin plug connectors. For lifting, a tubular strut is placed across the cab in a position immediately under the roof section (to prevent distortion during lifting), a lifting beam is attached to two lifting eyes temporarily fitted towards the front and rear on the centreline of the roof, and the whole unit can then be lifted off using an overhead crane.

One of the major recognisable features of the cab are the windows, and this is an area where there was a desire to match the original appearance as closely as possible. However, the latest safety regulations, particularly for high-speed running, required changes to the type of glass.

On the forward-facing windows, the original Peppercorn A1s were fitted with $^3/_8$in armoured

BELOW Back at Darlington and the cab is seen mounted on the frames. *(David Elliott/A1SLT)*

glass, but to meet current regulations, on *Tornado* this had to be 21.5mm laminated glass, with an anti-spall coating (of thin polycarbonate) on the inner surface. Whereas the original windows were fitted from the inside with a trim strip, also on the inside to retain them, it was decided that additional safety would be gained by fitting these from the outside so that the pressure on the rim of the glass acting against the frame would naturally resist the possibility of the window being forced inwards. This required careful alignment and precision drilling of the surround to maintain the correct appearance.

The side windows were originally ¼in glass in teak frames. Again, the regulations mandated that these be converted to safety glass. In this instance, as these windows were effectively the crew escape route if the locomotive ever tipped over on to its side and the tender was blocking the main access door, 9mm toughened glass was adopted, this being capable of being broken in emergency and forming less hazardous small crystals rather than sharp splinters of glass. The glazings were fitted into specially manufactured teak frames to maintain the correct appearance of the windows. The side screens were also made in toughened glass.

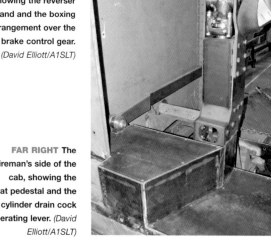

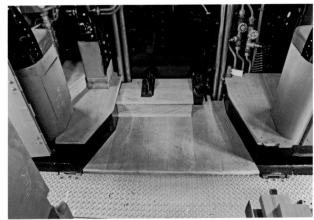

The original seats for the driver and fireman were bolted to frames directly fastened to the side structure of the cab. On *Tornado,* it was decided that the space under the seats could usefully house some of the electrical control equipment, and special pedestal cubicles were built in such a way that the seats could be fastened directly to the top of them. The seats themselves were kindly made and donated by Marshall Aerospace of Cambridge, although the vinyl cushion covering has since been replaced with leather for greater durability.

Platework and cladding

M ost of the platework consists of footplating above the wheels, and around the front of the smokebox area. This is all in the form of steel plates, bolted to brackets which in turn are bolted to the main frames. Small splashers are incorporated over the leading and intermediate coupled wheels. Whilst on the original Peppercorn A1s the construction would have been mostly riveted, the use of bolts simplified construction and has the added advantage of making removal easier, when needed for maintenance purposes.

The layout of the plates has been designed such that in areas where these will need to be removed regularly, they are in separate sections,

minimising the amount of work required to remove and refit these.

The other significant items of platework are the smoke deflectors which are made from 3mm steel plate reinforced with steel strip and angle, with beading around the top and front edges.

The boiler itself is insulated. The boiler and the firebox were fitted with crinoline rings secured at intervals around the boiler barrel. Ceramic fibre insulation material, sandwiched between two layers of aluminium foil, was placed around the boiler in between the crinoline rings. Steel cladding sheet, 2mm thick, was rolled to shape to fit around the barrel over

ABOVE LEFT General view of the cab nearing completion. Note the rocking firegrate operating lever. *(David Elliott/A1SLT)*

ABOVE The wooden flooring fitted to the cab. *(David Elliott/A1SLT)*

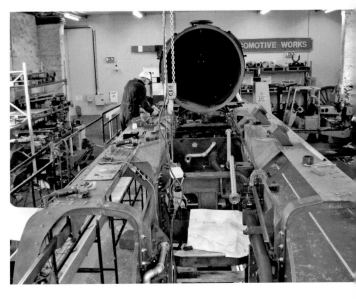

RIGHT The foot plating over the wheels is fitted to brackets attached to the main frames. *(David Elliott/A1SLT)*

the insulation, and securing bands of steel were fitted around the cladding – aligned with the crinoline rings to give a firm base, and secured with bolts underneath the boiler.

Lubrication

There are many moving parts on both the locomotive and tender, which have metal-to-metal surface contact, and to minimise wear, lubrication is essential. These fall into two categories – rotational parts (such as the wheelsets, and parts of the motion), and sliding parts (such as the piston and valve rods).

On the original Peppercorn A1s most of the lubrication was oil-based, with reservoirs fitted to most items. These were all topped up during the preparation of the locomotive for a run. Whilst oil is an excellent lubricant, it does have its problems, especially on items where it is not immediately visible, such as rotating parts. First, it is possible for small particles of dirt to get into the system, preventing complete lubrication and potentially causing wear, and secondly, water can get into the oil, leading to corrosion, again in parts which are not normally visible.

To minimise these problems it was decided to make some changes to the lubrication systems on *Tornado*, and in general terms the smaller rotating items are fitted with grease lubrication with grease nipples rather than oil reservoirs. This allows the new grease to be fed in, pushing the old grease out, until fresh grease is observed.

The sliding parts (which are for the most part exposed) cannot be dealt with in this way and the large motion bearings are still oil-lubricated, fed from a total of 14 oil reservoirs situated around the locomotive and tender (and in the case of motion bearings, by individual oil boxes built into each bearing). These all have to be topped up for each run (or daily when carrying out short runs on heritage lines). On a typical day, *Tornado* will use around two gallons of lubricating oil.

The fixed oil boxes provide gravity feeds, using a wick to syphon the oil up from the reservoir and drip this into a copper tube, which is positioned to drip the oil on to the

ABOVE The firebox cladding being cut to profile. *(David Elliott/A1SLT)*

RIGHT The boiler barrel cladding being rolled to shape. *(David Elliott/A1SLT)*

RIGHT The cladding was clamped around the crinoline rings. *(David Elliott/A1SLT)*

relevant part. Most of the boxes have several feed pipes serving various components.

There are four boxes around the cylinders at the front of the locomotive, which provide the lubrication to the piston rod and slidebars, the valve spindle and the valve spindle slidebars, and to the side bearer on the front bogie. These also serve the valve spindle covers on the front of the valve chests, and the inside cylinder motion. There is also an additional single-feed box serving the inside crosshead.

Two further boxes serve both sides of the front coupled axle, there are two feeds to the horn block liners, and a further two to the trunnion plates on the front axle boxes. Four more boxes serve each side of the centre and rear coupled axles, feeding the horn block liners, and a further two boxes serve each side of the Cartazzi axle – lubricating the slides in the axle boxes. A final single-feed box serves the tender handbrake screw mechanism.

In addition to the gravity-fed lubrication there are three further mechanical lubrication systems. The first of these is a six-feed lubricator, which pumps steam oil to lubricate the valve chest and cylinders. A single box, with preheated oil, has six off-takes – three inject the oil into the steam feed pipes to the valve chests, and three feed directly into the cylinders (the inlet valve hidden behind the small circular plate visible on the cylinder cladding). These all use live steam to atomise the oil (break it up into small droplets to ensure it is dispersed around the valves and cylinders).

Finally, each of the air pumps has a mechanical lubricator – each with two reservoirs. One of these contains steam oil which is fed into the steam cylinder of the compressor, and the other compressor oil, which is fed to the low-pressure air cylinder and the piston rod between the steam and air cylinders. Compressor lubricators are normally attached to, and are directly operated by each compressor. However, with *Tornado*'s air compressors being mounted in inaccessible locations between the frames, the decision was made to move the lubricators into the cab. They are now operated by small air servos actuated by air from one side of the low-pressure air piston on each pump.

ABOVE One of the oil lubrication reservoirs. *(Geoff Smith/A1SLT)*

LEFT The mechanical lubricator during dye penetration testing to assess its suitability for use. *(David Elliott/A1SLT)*

Sandboxes

In order to improve traction under wet and slippery conditions the A1s, in common with most locomotives, were fitted with sanding equipment. When needed, the sand was released on to the tracks immediately in front of the coupled wheels. *Tornado* has six sandboxes, two deliver sand to the rear of the centre set of coupled wheels when the

LEFT The mechanical lubricator. This is driven from the motion and provides lubrication for the valve chests and cylinders. *(Geoff Smith/A1SLT)*

RIGHT A sandbox
filler spout above the
side foot plating of
the locomotive. (Geoff
Smith/A1SLT)

FAR RIGHT The
arrangement of the
sand piping to one
wheel. Note the small
air pipe designed to
induce the sand into
the air stream. (Geoff
Smith/A1SLT)

locomotive is in reverse gear, while the other four deliver it on to the track ahead of the front and centre coupled wheels, when travelling in a forward direction. The middle and rear boxes are mounted on the inside of the frames, while the leading boxes are fitted on the outside, behind the slide bars. All are filled through access ports above the footplating.

The original Peppercorn A1s used steam sanders – small steam pipes adjacent to the sand outlet used a steam jet to induce sand down the sand pipe and forced it under the wheels. *Tornado* uses air sanders rather than steam, but the principle of induction remains. The sand pipe itself is shaped in such a way that friction prevents the sand falling under gravity alone, but as the air jet is turned on the induction effect is sufficient to overcome the frictional resistance and allow the sand to be entrained into the air stream. It is important that the sand itself should be dry, to prevent the system clogging and failing to work.

Electrical systems

The original Peppercorn A1s were equipped with limited electrical equipment, this being powered from a steam-driven 24v ac Stones turbo generator. This was rated at 350 watts and powered lighting systems only, which comprised four marker lights each on the front of the locomotive and the rear of the tender (individually switchable from the cab), as well as cab lighting (which itself comprised a general-purpose light on the roof of the cab, and specific lights for the boiler water level gauges, the pressure gauges, and the speedometer). Two inspection light sockets were also provided, one each side of the locomotive adjacent to the step in the running plate behind the main

cylinders. Two-pole sockets were provided on the underside of the locomotive and tender dragboxes to allow a cable to link the two.

The turbo generator had no electrical regulation, instead, relying on a centrifugal governor on the steam turbine to regulate the speed of the permanent magnet-type alternator – this giving a fairly constant ac output. The locomotive frames appear to have been used as the return conductor. There was no battery back-up, hence no power when the locomotive was not in steam, or in the event of a breakdown of the turbo generator.

Regulatory requirements for main line running have changed significantly since the 1950s, and this includes a substantial increase in the amount of electrical equipment on board, particularly in the form of current, and anticipated, safety-related signalling, monitoring and recording equipment. Not only has the amount of equipment increased, but there is also now a need for the electrical supply to it to be secure and uninterruptible.

This led to a need to completely redesign the electrical systems, and from the outset, the whole design concept was built around reliability (minimising the risk of failure, by close attention to details such as cable routing, minimising and careful selection of cable connectors, build quality etc), and maintainability (facilitating fault detection, and access for maintenance). This had to be implemented within the testing environment of a moving steam locomotive, with the attendant problems of heat, dirt and dust, and vibration. In addition, all designs had to comply with the relevant Railway Group Standards for electrical and fire safety, EMC (electromagnetic compatibility), earthing, shock, and vibration.

To put this into context, the installation

uses over 2.4 miles of wiring, and some 9,000 electrical and electronic components. There are 230 separate wiring runs, involving 52 heavy-duty military connectors, and 500 separate connections.

It was decided that there would be no mains voltages on board the locomotive or tender – all electrical systems would be extra low voltage (ELV), defined as no more than 50v ac or 120v ripple-free dc. In addition, the metal of the locomotive itself would no longer be used as the main return path, separate neutral cables would be installed throughout.

Following the principles of supply security, the locomotive would have three power sources, each feeding two identical and independent battery systems, the batteries being charged by the generators whenever these were operational. The three supplies are:

Steam turbo generator – this provides power when the locomotive is in steam and stationary. A suitable unit was sourced from DB Meiningen, incorporating a Bosch alternator, generating 32 amps at 28v. This is built into a purpose-designed housing, located behind the right-hand smoke deflector, to simulate that used on the original Peppercorn A1s.

Tender alternator – this is mechanically driven from the rear tender axle, and provides power whenever the locomotive is in motion. A suitable alternator was sourced from an old Mk 1 Royal Mail sorting van, refurbished, and fitted to a custom-designed mounting on the tender frame. The drive is taken from the tender axle via adjustable Fenner V belts. This unit has a nominal output of 200 amps at 24v, and is fitted with a regulator/rectifier box to convert the output to dc.

Shore power – in addition to the above self-contained generator systems, there are also connectors on each battery box to allow 24v dc power to be plugged in when the locomotive is idle on shed. These enable direct feeds to the Input/Output (I/O) panels and to the battery chargers.

The electrical services systems on the locomotive are split into two – the essential services system, and the auxiliary services

ABOVE The steam turbo generator, obtained from DB Meiningen. *(David Elliott/A1SLT)*

LEFT The tender alternator mounted on the tender frames. *(Nigel Facer/A1SLT)*

system, each of these having its own battery supply, located in purpose-built boxes mounted to the outer rear frames of the locomotive, in front of the crew footsteps. The one on the driver's side houses the essential services battery, and the one on the fireman's side the auxiliary services battery. Each battery system has a rating of 65AH at 24v, and each is fitted with its own charger and protection for charge over-voltage and discharge over-current.

Each of the battery boxes has its own I/O

BELOW The assembled shore power unit on test prior to being fitted. *(Rob Morland/A1SLT)*

ABOVE The essential services and auxiliary services input/output (I/O) panels prior to installation. *(Rob Morland/A1SLT)*

ABOVE RIGHT The auxiliary services I/O panel and switch panel during construction and testing. *(Rob Morland/A1SLT)*

panel, with miniature circuit breakers (MCBs) for input protection. There is automatic selection of the charging source – steam turbo generator, tender alternator, or shore power. In addition, each box is configured so that it can supply either the essential services or auxiliary services system, or both, giving resilience in the event of failure of either of the battery supplies.

The essential services supply serves the following loads:

- Automatic Warning System (AWS)/Train Protection Warning System (TPWS)
- On Train Monitoring and Recording System (OTMR)
- Cab Radio(s)

- Essential cab instrument lighting
- Essential head lamps, tail lamps and marker lamps for both the locomotive and tender
- Provision for future essential equipment

The auxiliary services supply serves the following loads:

- Cab space lighting
- Auxiliary cab instrument lighting
- Locomotive injector overflow lights
- Locomotive frame inspection lights/coal bunker lights
- Locomotive inspection lamp plug-in points
- Non-essential instrumentation, etc
- Provision for future auxiliary equipment

RIGHT The essential services I/O panel installed under the driver's seat. *(Rob Morland/A1SLT)*

FAR RIGHT The auxiliary services I/O panel and the On Train Monitoring Recorder (OTMR) installed under the fireman's seat. *(Rob Morland/A1SLT)*

Normally, each battery box feeds its own I/O panel, the essential services panel being located under the driver's seat, and the auxiliary services panel being located under the fireman's seat. Each panel is fitted with MCB-protected circuits for each load, and each panel includes spare circuits for future use.

The battery charging systems have been designed to comply with the battery manufacturer's specifications and can be safely left on a float charge indefinitely, without damage to the batteries.

The power system operates on a nominal 24v dc with a negative earth, and equipotential bonding points are provided from each system to the locomotive frames.

The main I/O panels supply control panels located on the cab roof above the crew seats. Again, the essential services panel is on the driver's side, and the auxiliary services panel on the fireman's side. The control panels provide switches for all the loads which require routine operation by the crew, together with indicator lamps to advise the crew in the event of a fault developing either in the supply to a circuit, or the failure of a lamp in that circuit.

The essential services loads are serviced as follows:

AWS/TPWS – these are safety systems both designed to reduce the incidence of passing signals at danger. The AWS was first introduced in 1956. On the track, the equipment consisted of two magnets, the first being a permanent magnet and the second an electro-magnet, located approximately 197 yards (180m) before the signal. If the signal ahead was anything other than green, the electro-magnet was de-energised and the signal given to the driver was a horn mounted in the cab, with automatic application of the brakes if the cancel handle was not depressed. For a green signal, the electro-magnet was energised and this cancelled the horn and instead rang a bell in the cab. It also cancelled the brake application procedure. In both cases, the indicator changed from black/yellow segments to black only segments on passing the permanent magnet.

LEFT Paul Depledge wiring the auxiliary services I/O Panel. *(Rob Morland/A1SLT)*

BELOW LEFT Paul Depledge wiring the auxiliary services battery back-up power supply, located on the outside of the frames, below the cab on the fireman's side. *(Rob Morland/A1SLT)*

BELOW The essential services lighting switch panel installed on the cab roof (showing temporary labels at this stage). *(Rob Morland/A1SLT)*

RIGHT The National Radio Network (NRN) cab radio. *(Rob Morland/A1SLT)*

If the signal was anything other than green, the depression of the cancel handle turned the display back to yellow/black. In the case of a green signal, the display remained all black until another signal was approached which showed an aspect other than green.

The detector fitted to the locomotive enclosed an armature, which was pulled

RIGHT Paul Depledge installs the NRN radio antenna on top of the tender. *(Rob Morland/A1SLT)*

RIGHT Traction Inspector Malcolm Hall uses the NRN Radio. *(Rob Morland/A1SLT)*

one way by the permanent magnet and then immediately pushed in the opposite direction by the electro-magnet, if the signal being approached was green.

The TPWS was a later development of, but not a replacement for, the AWS. It is based on two pairs of loops set between the rails, one some distance from the signal and one closer to the signal (or other obstruction, as TPWS is also used in other applications such as the approach to buffer stops). Each pair of loops consists of an arming loop and a trigger loop, and the first pair measure the speed of the train as it approaches the obstruction – if it is going too fast to be able to stop the brakes are automatically applied irrespective of any action or inaction on the part of the driver. If it passes the first pair of loops successfully then it is tested again at the second pair of loops, if the signal is at danger. Again, if it is obviously going to pass the signal then the brakes are automatically applied.

The detection unit for these systems on *Tornado* is located under the driver's seat, and is a standard unit supplied by Thales. Three sensor antennas are installed, two on the locomotive and one on the tender. These are switched by a changeover switch on the reverser, to select the locomotive or tender, depending on which way the locomotive is running.

OTMR – the On Train Monitoring and Recording Equipment operates in a similar way to the well-known 'black box' flight recorder on aircraft. It is designed to record multiple channel data on the operation of the locomotive including locomotive speed, steam chest pressure, number and type of brake applications, brake pipe pressure, vacuum value in inches of mercury, AWS and TPWS operations, for later download and analysis. It is designed to withstand damage (e.g. in the event of an accident). The unit fitted to *Tornado* was supplied by Arrowvale, and is located under the fireman's seat.

Cab Radio – this is fitted in a purpose-built steel cabinet fixed to the tender on the fireman's side of the cab. This houses a Philips radio (and associated power supply) operating on the National Radio Network and provides two-way communication between the locomotive and land-based systems. The cabinet is designed to allow the future fitting of a GSM-R radio or any other radio system in the future.

Essential instrument lighting – this is provided by a cluster of narrow angle light emitting diodes (LEDs) installed in a small light box mounted on the cab roof on the driver's side. These are individually directed to illuminate various instruments including the driver's side gauges, speedometer, reverser position indicator, and the left-hand water gauge. The driver can control the intensity of the lighting by a small potentiometer located on the light box. A single LED on an umbilical cord is used to provide lighting for the essential services control panel, and again, the intensity of this is adjustable by potentiometer.

Marker and tail lamps – these follow the original design of marker lamps with four square lamps mounted on each of the front of the locomotive and the rear of the tender. Although the original-type housings are used, these are fitted with LEDs, designed to meet the GM/RT standard for marker lamps for main line use.

In addition, the lower left and right-hand markers on both the locomotive and tender are also fitted with red LEDs and beam splitters, selectable from the essential services control panel, to allow their use as marker/tail lamps, depending on the direction of travel of the locomotive, and again, in accordance with the required standard.

Headlamps – steam locomotives operating on the National Network are required to have a high-intensity forward-facing lamp on the bufferbeam. This is normally provided by a portable headlamp which is permitted at speeds up to 75mph.

Tornado is provided with semi-permanent headlamps. Although using replica oil lamp housings, these are to a new design using an array of high-intensity LEDs. The lamps are provided with sensors alerting the crew via the essential services control panel in the event of part of an array failing. Once again, these lamps have been designed to meet the intensity requirements of the current standard for day (left-hand side lamp when viewed from the front)

LEFT The LED instrument lighting boxes under test prior to fitting. *(Rob Morland/A1SLT)*

BELOW Four front marker lamps in place. The steam turbo alternator is behind the left-hand smoke deflector. *(Rob Morland/A1SLT)*

and night (right-hand lamp when viewed from the front), running at speeds in excess of 60mph.

The lamps are also provided with a hazard-warning mode, controlled by a push switch on the essential services control panel, which flashes the lamps at around 40 cycles per minute. Illumination in the switch alerts the crew that this mode is selected.

The auxiliary services loads are serviced as follows:

Cab space lighting – two bulkhead fittings are positioned on the cab roof. To maintain an authentic appearance with the original Peppercorn A1s these are fitted with incandescent lamps.

Auxiliary instrument lighting – this is of a similar design to the essential instrument lighting, via a light box fitted to the cab roof on the fireman's side. In this instance, the narrow angle LEDs illuminate the boiler pressure gauge and the right-hand water gauge. Light intensity is again controlled by a small potentiometer.

LED lighting is also provided for the tender instruments, the coal shovelling plate, and, again, mirroring the arrangement for the essential services panel, via a single LED on an umbilical to illuminate the auxiliary services control panel.

Locomotive injector overflow lights – these are provided by long-life LEDs, in fittings mounted on the underside of the lower cab steps on both the driver's and fireman's sides of the locomotive. These lamps have a useful secondary function in providing illumination of the ground for the crew when alighting from the locomotive at night.

Locomotive frame inspection lights/coal bunker lights – to assist with inspection/maintenance, six bulkhead fittings complete

with long-life LED lamps have been fitted
between the frames of the locomotive and
tender. In addition, LED strip lighting has been
installed under the running plates down each
side of the locomotive to provide illumination of
the motion and wheels.

Two bulkhead fittings are provided in the
coal bunker to assist with the movement of coal
at night.

Locomotive inspection lamp plug-in points
– a number of sockets are provided around the
locomotive and tender to allow the plugging in
of hand lamps when inspecting or working on
the locomotive.

All wiring on the locomotive and tender
is fitted within conduit or trunking to give
mechanical protection, and to permit the easy
replacement of cables in the event of failure. The
cab systems are designed such, that if removal
of the cab is required, say in the case of major
overhaul, most will remain within it during its
removal. Connections are provided so that all
cables which service loads outside the cab are
fitted with connectors at the cab/frame interface,
to allow for easy disconnection/reconnection.

Connections between the locomotive and
tender are via umbilical leads with heavy-duty
connectors on either end. Each circuit and its
associated cabling is identified by a four-digit
alpha-numeric code, using the E designation
for essential services, and the A designation for
auxiliary services.

There has been a lot of discussion in the
railway press about the potential introduction
on Network Rail of the European Rail Traffic
Management System (ERTMS), and the effect
that this will have on steam traction. ERTMS
has been developed as a Europe-wide safety
system which will allow locomotives from
different countries within the EU to operate on
the network in any other country to the same
safety standards. It would also increase the
capacity of the network by allowing trains to
follow each other more closely at high speeds
but in full safety. In principle, the position of any
train is known, using positioning technology,
and its speed is automatically adjusted
according to network conditions ahead of it.

The system is still in development and
testing, and each country has still to decide on

ABOVE **Connecting
cables to the
heavy duty military
connector.** (Rob
Morland/A1SLT)

LEFT **The locomotive-
to-tender umbilicals
wired and ready for
use.** (Rob Morland/
A1SLT)

LEFT **Paul Depledge
with the locomotive-
to-tender umbilical
electrical connections
in place.** (Rob Morland/
A1SLT)

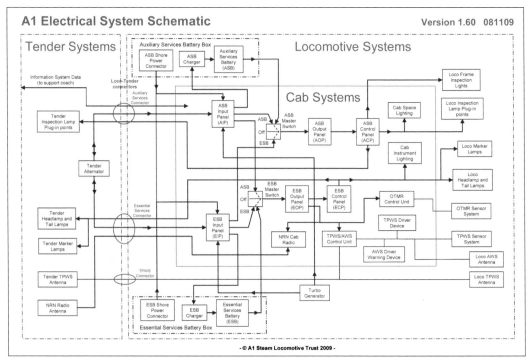

Tender Systems | Locomotive Systems

Cab Systems

the best method of application, but if, or more probably, when, adopted in the UK, this will require the addition of considerable additional equipment on each locomotive. *Tornado* has been designed to have the electrical capacity required to cope with this when it is introduced.

The electrical schematic is shown above.

Tender

The original Peppercorn A1s were fitted with tenders with a coal capacity of 9 tons and a water capacity of 5,000 gallons. They were also fitted with water scoops to take advantage of en-route water top-ups, using the water troughs which were fitted between the rails at regular intervals on main line routes. The scoop was lowered into the water on approach to the troughs and the forward speed of the locomotive forced the water up into the tender. In addition to the troughs, most stations had a water top-up facility via a mains-fed tank and bag hose which was operated by the footplate crew.

With the demise of steam, both the water troughs and the station watering facilities were

removed, limiting the options for water filling of heritage steam locomotives. With a water consumption on *Tornado* of some 30 to 32 gallons per mile when in normal main line operation, although quite economical for a large steam locomotive, all main line runs require water top-up stops – usually arranged by road tanker at a convenient stopping point.

With the scoop and associated control gear no longer required, the opportunity was taken to redesign the tender to increase the water capacity to some 6,270 gallons, whilst at the same time reducing the coal capacity to 7.5 tons. This would assist in service by reducing the frequency of water stops, but was a substantial change in design and required the approval of the Regulatory Authorities before it could proceed. Although the increase in the weight of water (approaching 6 tonnes) was offset in part by the reduction in the weight of coal, and the removal of the water scoop gear, water is a fluid and moves about due to the motion of the locomotive. It was therefore necessary to prove that the redesign was stable, with a low enough centre of gravity to avoid any tendency to tip over in operation, particularly when the water tank was full.

To prevent instability due to the movement of water in the tank, the original Peppercorn A1s had a series of baffles in the tender. It was necessary to include and improve these baffles inside *Tornado's* tender, both across it and longitudinally, due to the increased water capacity. There are two longitudinal baffles, and six transverse baffles, which are carefully designed to allow movement of water between the compartments, but at a rate which does not affect the stability of the tender when subject to sudden actions at speed (e.g. emergency braking). To allow access for inspection, cleaning and maintenance, man hole doors are provided in the baffle plates at relevant points, to ensure that all parts of the tank are accessible internally. The base of the tank has a lower 'well' section from which the feed water is drawn off to the locomotive. The baffle plates do not dip into the well, as the amount of water remaining at this point would not be sufficient to cause any problems due to water movement.

In order to carry out the necessary calculations for approval it was important to understand the internal construction of the original tank and its baffles, which is not an easy task from a two-dimensional drawing. To overcome this problem, a scale three-dimensional cardboard cut away model was built, prior to finalisation of the plans.

Another design issue related to the wheels. Most of the Peppercorn A1 ran with disc-wheel tenders – the wheel effectively being solid. As all modern traction runs with disc wheels, the 'standards' for these have been developed over the years and are now very prescriptive. Conversely, spoked wheels, which disappeared with the demise of steam traction, are not subject to any specific 'standard'. The market was researched to try to find a manufacturer who could provide disc wheels of the size required, and although an overseas supplier was eventually identified (who had made the wheels for the Channel Tunnel Euroshuttle locomotives), they had a minimum order of 200 wheels! This was considered slightly excessive, against a current requirement of eight, and attention turned to alternatives. Careful research eventually identified that a small number of

the original Peppercorn A1s ran with spoked wheels, and it was decided that this was the practical way forward.

The wheels, in common with the locomotive wheels, were cast by William Cook Cast Products, who also carried out the preliminary machining. The forged axles were supplied by Ring Rollers in South Africa, and the fitting of the wheels to the axles, final machining, and fitting and profiling of tyres was carried out by Ian Riley Engineering. The tender chassis was constructed by I. D. Howitt of Crofton, Wakefield.

The basic frames were supplied flame cut, and the first task was to mill these to the correct size and shape for the subsequent attachments. To ensure accuracy the frames were clamped together so that the machining took place on both sides at the same time, as if they were one piece of metal. The most critical area of this machining was to accommodate the horn blocks, in which the wheel axle boxes sit, and where dimensions are critical to ensure that the axle boxes can slide in the vertical plane, but with minimum movement backwards and forwards.

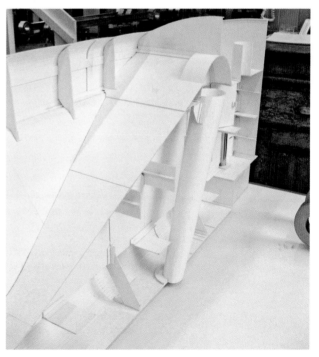

BELOW Part of the cardboard model of the cut-away tender showing the original water scoop.
(David Elliott/A1SLT)

LEFT The tender side frame with hornblocks and brackets attached, at Ian Howitt's Crofton Works. *(Ian Howitt/ A1SLT)*

ABOVE The inverted tender frames, positioned for the attachment of the frame stretchers. *(Ian Howitt/A1SLT)*

BELOW Stretchers being attached to the tender frames at Ian Howitt's Crofton Works. *(Ian Howitt/ A1SLT)*

LEFT The tender rear dragbox fabrication, which was fitted between the rear of the frames. *(Ian Howitt/ A1SLT)*

RIGHT The tender front dragbox held in position and ready for the drilling of the fixing holes. *(Ian Howitt/A1SLT)*

Once machining was complete it was then necessary to drill the bolt holes to attach the horn blocks and other brackets which are bolted to the frames. These include the front and rear dragboxes (which form the attachments to the locomotive and train respectively), the bufferbeam brackets, frame stretchers (the steelwork which sits between the frames, and effectively maintains the spacing), spring hangers, brake gear hangers, and the brackets for the tank fabrication, which sits on top of the frames. On the original Peppercorn A1s, most of these would have been riveted, but in this instance it was decided that, in order to facilitate subsequent maintenance, all attachments should be made using fitted bolts, which are a precision fit to eliminate any potential for small movements of the bolt within the hole. This involves pilot drilling the hole, followed by the main drill slightly under size, and then using a reamer to produce the final hole dimension. As an added security self-locking nuts were used throughout.

To facilitate this operation, the whole of this work was carried out with the frames in an 'upside-down' position. The next stage was to drill the holes in the tank base to allow fixing to the frames.

The tank itself was fabricated by North View Engineering in Darlington. Bearing in mind the decision to use bolts to secure the tank to the frame, it was considered it would not be possible to drill the tank base sufficiently accurately for the fixings, by using measurements alone. Once the tank base fabrication was completed it was therefore taken down to Wakefield to be mated up with the frames for drilling. To achieve this it was necessary to turn the frames upright – which required the use of a heavy-duty crane, and some careful manoeuvring.

Once aligned, the tank base was drilled, and to facilitate subsequent connection to the frame (once fabrication was complete), sealed captive nuts were welded to the inside of the base, ready to accept the bolts inserted from the underside.

The tank base was then returned to North View Engineering for completion of the fabrication, and the frames were inverted once again to complete the fabrication and erection

ABOVE The tender frames being turned to the upright orientation ready for a trial fit of the baseplate of the tender top section. *(Nigel Facer/A1SLT)*

BELOW The upright frames were moved back into the workshop. *(Nigel Facer/A1SLT)*

LEFT The tender base after drilling and fitting of captive nuts, loaded and ready to return to North View Engineering in Darlington. *(Ian Howitt/A1SLT)*

process. At this point, a further problem occurred with the design of the brake gear. The tender brake gear is not compensated, as it comprises a series of linked rods driven from a single point of application, which brings all the blocks into contact with the wheels at the same time. The brakes on the original A1s were steam operated, on *Tornado* they would be air operated. Although the pressures exerted would be roughly the same, as this was a design change, approval was required, and this in turn needed calculations to show that the gear was up to the stresses which would be created.

The calculations revealed that in the worst case scenario there was a potential problem when new brake blocks are fitted. As brake blocks are unmachined castings their dimensions vary slightly, which can lead to one brake block coming into contact with its wheel ahead of the others – continued pressure to bring the other blocks into contact with the wheel would result in heavy stresses in the one applied first, and this could cause distortion of the brake hanger. It would appear that the design of this brake gear had its history way back with the smaller tenders of the Great

Northern era. This had been adapted to suit larger tenders during the Gresley/Thompson eras and on into the Peppercorn A1s without ever being modified. Thus the 'potential' weakness had always existed, but had never proved to be a problem. Nevertheless, this required a redesign of the hanger itself into a significantly more robust component to satisfy the Regulatory Authorities. Subsequent experience in operation has shown this was very successful.

The tank fabrication comprised the welding together of a lot of steel plate, to form the shape of the tender, and the internal baffling required.

LEFT The front of the tender during fabrication, showing the coal storage area. *(David Elliott/A1SLT)*

BELOW LEFT The rear of the tender during fabrication, showing the right-hand tender side and the water tank longitudinal baffles. *(David Elliott/A1SLT)*

BELOW The tender top nears completion at North View Engineering, Darlington. *(David Elliott/A1SLT)*

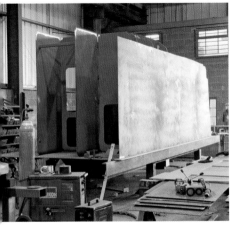

component. Although fitted on the outside of the wheels themselves, these required special pressing on to the axles and this work was carried out in a single day by the Trust team, with consultancy supervision being provided by the Timken representative free of charge under a sponsorship arrangement. The bearing cartridge is cylindrical, and inverted 'U'-shaped axle boxes were cast and machined, to correspond to the bearing. Horn stays were then bolted to the main frame plates underneath the axles to retain them in place if the tender needed to be lifted.

The next stage was to fit the various components which had to be located inside the chassis, most notably the pipework. With a significant amount of extra equipment under the tender – air tanks, electrical generator, air and vacuum pipes, and electrical conduits – space was very limited, particularly so as the tender tank well drops down between the frames, making access to the upper part of the inner frames almost impossible with the tank in place.

Once this work was completed the tank itself could be lowered on to the frames and bolted into place. Prior to delivery, the tank was filled with water to test for leaks, but only minor ones were found and these were soon solved by rewelding the affected areas.

With the tank now in place, work could

ABOVE The completed tender frames were loaded on to a wagon at Ian Howitt's Crofton Works ready for transport to Hopetown Works. *(Nigel Facer/A1SLT)*

Once the fabrication was complete, the three major components – frames and associated equipment, tank, and wheelsets – were moved to Darlington ready for final assembly to take place.

The first task was to get the frames on to the wheels to give a rolling chassis, but before this could be done the wheel bearings and axle boxes had to be fitted, so that these could be located in the horn guides in the frames. The tender, in common with the locomotive, uses Timken roller bearings, but in this instance they are 'cartridge units', supplied as a complete assembled and pre-greased

RIGHT The tender wheelsets in Hopetown Works, ready for the arrival of the other two main sections of the tender. *(David Elliott/A1SLT)*

continue on the 'finishing' parts of the tender, adding the buffers, various valves and connections, and adding the filler pipes. As stated earlier, water fill-ups are usually from road tankers or fire hydrants – these can be at either side of the locomotive, depending on location. It was therefore decided to fit filler pipes on both sides of the tender, and in order to speed re-watering (the time taken being an issue on heritage locomotives) to fit two connections on each side. This allows two hoses to be fitted, irrespective of which side of the tender the tankers are positioned, and halves the watering time.

Steps and a vertical hand rail were fitted to the rear of the tender to facilitate access to the top of the tank, but each step had to be fitted with an obstructer in the form of a taper block, bolted through the step itself, to prevent their use when under overhead electric wires. These have to be unbolted to allow the steps to be used when the locomotive is in a suitable location.

The cab end of the tender was fitted with cubicles, the one on the driver's side houses the air-brake control equipment, while the one on the fireman's side houses the NRN radio, as detailed in the electrical section. This box has other compartments used for general storage, and also fitted to this end of the tender is a mobile 'phone charger – a sign of the times!

TOP The tender base was lowered on to the wheelsets, with the wheel bearings guided into the horns on the frames. *(David Elliott/A1SLT)*

ABOVE With the frames and wheelsets now joined, the assembly was placed back inside Hopetown Works to await arrival of the tender top. *(David Elliott/A1SLT)*

LEFT The tender top was lowered on to the frames at Hopetown Works. *(David Elliott/A1SLT)*

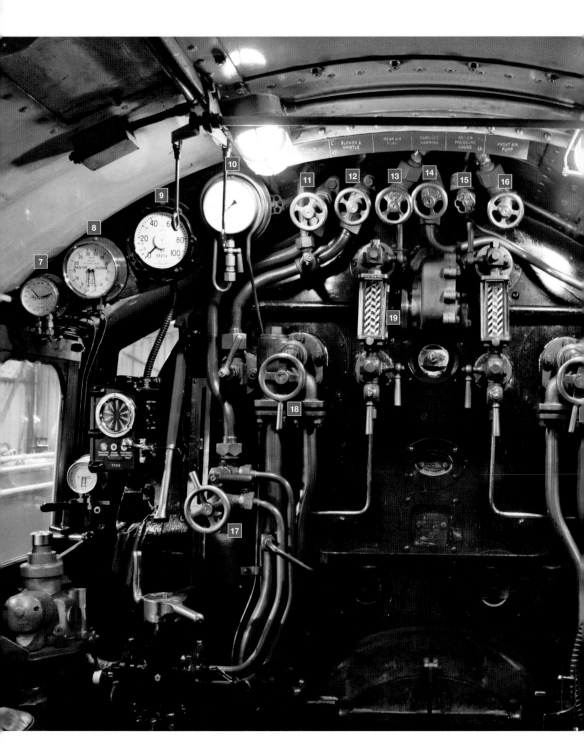

LEFT The boiler backhead control layout.

(Graham Nicholas/A1SLT)

ABOVE The driver's side control layout.

(Graham Nicholas/A1SLT)

BELOW The fireman's side control layout.

(Graham Nicholas/A1SLT)

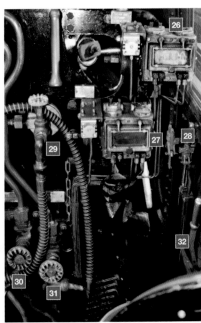

Controls

The main operating controls are all contained within the cab, the layout of which is shown here.

Key to photographs

1. Driver's automatic air brake valve
2. Straight air brake valve
3. Dual air pressure gauge 2 showing air brake pipe and locomotive brake air pressures
4. Reversing handle
5. Regulator handle
6. Combined AWS/TPWS indicator/acknowledgement unit
7. Dual air pressure gauge 1 showing main reservoir and train reservoir pipe air pressures
8. Dual vacuum brake gauge showing vacuum brake pipe and vacuum control reservoir pressures
9. Speedometer
10. Steam chest pressure gauge
11. Steam to vacuum ejector valve
12. Steam to blower and whistle valve
13. Steam to rear air pump valve
14. Steam to turbo alternator valve
15. Steam to boiler pressure gauge valve
16. Steam to front air pump valve
17. Large and small ejector valves
18. Combined live steam injector steam and delivery valve
19. Boiler gauge glass assemblies
20. Combined exhaust steam injector steam and delivery valve
21. Standard whistle operating lever
22. Auxiliary whistle operating lever
23. Train heating pressure gauge
24. Steam pressure gauge
25. Steam to train heating valve
26. Rear air pump lubricator
27. Front air pump lubricator
28. Brake reservoir release valves
29. Cab water slacker pipe and valve
30. Ashpan water sprinkler valve
31. Tender water sprinkler valve
32. Cylinder drain cock handle

Wherever possible, the aim was to replicate the control layout of the original Peppercorn A1s. However, the essential modifications to allow *Tornado* to operate on the main line inevitably generated a requirement for additional indicators and controls. These had to be integrated in the original layout, but in a way which allowed the driver/fireman easy access.

Items 1, 2, 3, 7 and 8 relate to the air-braking system which was a new fitment on *Tornado*. Control levers 1 and 2 allow the driver to independently operate the automatic air brake and the straight air brake, the latter acting on the locomotive only.

Item 4 is the reverser. This is the largest piece of equipment in the cab, and consists of a vertical casting mounted off the floor, which houses the main screw shaft, which is operated by the handle mounted on top of the column. The top section, which is a fabrication, also houses the locking system which prevents any accidental movement of the reverser handle. The handle is attached to the top of a long nut which rotates on a screw attached at the bottom to a bell crank that transits the rotation of the handle to a backwards and forwards movement of the reversing rod itself, on the driver's side of the engine. This in turn translates to the up and down movement of the radius rod within the expansion link, controlling the direction of motion and the degree of cut off. The movement of the reversing rod is transmitted to the fireman's side cylinder motion gear via the weighshaft, which is a rod connecting the three sets of motion laterally across the locomotive, through the frames. The middle cylinder, being further forward than the outside two, has a separate weighshaft connected to the main weighshaft by two cranks and the inside reversing rod.

Item 5 is the regulator, which operates the main steam valve, admitting steam to the cylinders. This is a proportional control, in that the further it is moved, the greater the valve opening, and the greater the volume of steam available at the cylinders. There is a second regulator handle on the fireman's side for use in emergencies.

Item 6 is the combined AWS/TPWS warning and acknowledgement unit. The original A1s were fitted during their lifetime with the AWS unit, but the combined unit is new. The detail of the way these units operate is given in the Electrical systems section.

Item 8 is the dual vacuum brake gauge – one pointer shows the vacuum pressure in the vacuum brake pipe, the other the vacuum pressure in the control reservoir associated with the locomotive brake distributor.

Item 9 is the speedometer, worked by a generator attached to a return crank off the crankpin of rear coupled wheelset on the driver's side of the engine. This measures the number of wheel rotations, which is translated to a linear speed for display on the speedometer.

Item 10 is the steam chest pressure gauge which measures the steam pressure in the delivery pipe to the cylinders (after the regulator valve). This allows the driver to see what is happening in terms of steam entering the cylinders, and hopefully reduces slipping by allowing a controlled build-up of pressure.

Items 11–16 are the steam valves off the steam stand. These are self-explanatory from their descriptions and are described in more detail in the Pipework section.

Item 17 is the ejector valve, which controls the steam supplies to the large and small vacuum ejector nozzles.

Items 18 and 20 are the combined injector steam and delivery valves which control the flow of steam to the injectors, and contain the water delivery (clack) valves. These are equipped with emergency shut off valves which can be operated to prevent the boiler emptying through an injector overflow if a clack valve fails or leaks excessively.

Item 19 is the boiler gauge glasses. These are duplicate gauges which indicate the level of water in the boiler and are equipped with shut off and drain valves which can be operated if a gauge glass bursts, and are also used in sequence to test that the gauges are functioning correctly.

Items 21 and 22 are the two whistle controls and are repeated on the fireman's side, the inner handles being for the standard single-tone original A1 whistle, the outer handles controlling the auxiliary whistle pad which carries the chime whistle. Obviously, on the original Peppercorn A1s only one of these would have been present.

Items 23 and 24 are further gauges to indicate the pressure of the steam in the train heating system, and the main boiler pressure.

Item 25 is the valve controlling the steam supply to the carriage steam heating

Items 26 and 27 are new items on *Tornado*, being the mechanical lubricators for the two steam-driven air compressors. There are two reservoirs on each lubricator, one serving the air side of the pump with light, air-compressor, oil and the other serving the steam side of the pump with steam oil.

Item 28 is the brake reservoir release valves, which are used to empty the brake distributor air and vacuum control reservoirs.

Items 29, 30 and 31 are water valves controlling the water supplies to the cab slacker pipe and tender sprinkler (designed to reduce dust from dry coal), and the ashpan sprinkler, used from time to time to keep the ashpan platework cool to prevent it burning away or distorting.

Finally, item 32 is the cylinder drain cock operating lever, which releases any condensed water in the cylinders at the commencement of a move.

Painting and lining

The most important reason for painting any metal (but particularly ferrous metals) is to prevent oxidation (rusting) and corrosion. However, it is also a key component of presentation – corrosion can be prevented by paint of any colour, but careful selection of colour can enhance any object, and particularly so in the case of a large steam locomotive.

The original Peppercorn A1s appeared in three different colours over their lifetime: apple green, blue and Brunswick green, although with the development of the BR logo over the years, the Brunswick green version appeared with two different emblems on the tender – pre-1956 these had the lion-over-the-wheel emblem, while post-1956 this changed to the lion holding a wheel over the crown crest.

There is little that provokes more controversy among steam locomotive enthusiasts than liveries. This is partly driven by an overriding passion in some circles for absolute historical accuracy, but the bulk of the debate really centres around age and what enthusiasts remember. Those who are able to remember the pre-1948 nationalisation era tend to favour the liveries used by the 'Big Four' (the London & North Eastern Railway (LNER), the London Midland & Scottish Railway (LMS), the Great Western Railway (GWR), and the Southern Railway (SR)).

Right from the outset, it was always stated that *Tornado* was not to be reproduction of any of the 49 locomotives originally built, but to be the 50th member of the A1 class, incorporating a number of modifications to suit current main line running requirements – which meant that strict historical accuracy was not an issue. However, there was always a desire to follow as closely as practical the appearance of the original Peppercorn A1s, and it was agreed that *Tornado* would appear in each of the liveries carried by the originals over its first ten-year boiler certificate. For a number of reasons, the main one being the desire of the Trust's Honorary President, Dorothy Mather (widow of the designer of the original A1s, Arthur Peppercorn), to see *Tornado* as the original Peppercorn A1s first appeared from the workshops. It was therefore considered appropriate that the first livery carried should be apple green.

The first few months of *Tornado*'s active life were to be spent undergoing tests and trials in order to secure an operating certificate to be able to haul public trains on the main line network. This was always likely to involve engineering work and adjustments during this period, and to avoid potential damage to its finished paintwork, it was decided to carry out the initial painting in a grey 'undercoat' format. In practice, this was not undercoat, but a satin gloss which had a higher wear resistance.

Painting was carried out at the Darlington works where *Tornado* was built, and represented a particular challenge in that, in order to meet deadline dates, painting had to be carried out while final engineering work was still being completed. Rubbing down and preparation work could be carried out while other work was in progress, but in order to avoid dust damaging the wet paint, the actual application of the paint had to be left until evenings, and often late into the night. The

ABOVE The insides of the main frames were painted quite early on in the construction for reasons of accessibility. *(David Elliott/A1SLT)*

RIGHT Ian Matthews rubs down part of the outside of the frames ready for painting. *(David Elliott/A1SLT)*

painting was carried out by Ian Matthews and his son Daniel.

The initial part of the process was to degrease the surface using a solvent panel wipe, which is a special preparation used widely in the vehicle industry for removing grease, oils and silicones from bare metal or previously painted surfaces. Following this, the surface had to be rubbed down with an abrasive to roughen it. Although the roughening effect is barely visible to the naked eye, it is sufficient to provide a 'key' for the paint. Another wipe down to remove any particulate matter, and then one coat of oil-based primer/filler was applied by brush.

Any non-ferrous metal parts to be painted had to be initially coated with an etch primer. This is a special primer which contains a weak acid that attacks the surface of the metal, and again, provides a key for the subsequent paints. The etch primer is applied very sparsely, so much so, that it is often difficult to see that it has been applied at all.

Once the primer was dry four coats of an oil-based undercoat were applied to achieve a good protective layer. Timing was important with each coat needing to be applied within a period of 24 to 36 hours after the preceding coat. This allows sufficient time for the evaporation of most of the solvents in the previous coat, while leaving it still soft enough to bond with the following coat, allowing the strength to be attained. At this point, it was necessary to leave the undercoat for around two weeks to harden fully before the process could continue.

The hardened surface was then rubbed down again with a fine abrasive paper to smooth the surface and remove any blemishes, followed by a wipe down with the solvent panel wipe to remove any dirt and dust particles. A final coat of undercoat, mixed with an additive, was then applied, followed by two coats of the grey satin gloss, to produce the appearance of the locomotive for its running-in period.

Whilst the grey finish was used for purely practical reasons, the appearance immediately found favour with many enthusiasts as it undertook some four months' of tests and trials, initially on the preserved Great Central Railway in Leicestershire, and then on the main line.

ABOVE Ian paints the firebox cladding. (Rob Morland/A1SLT)

With the trials successfully completed, attention could turn to producing its first main line running livery – apple green. The locomotive was moved to the National Railway Museum at York, who kindly made their paint shop available. However, the initial cleaning and rubbing down (using a 320-grade abrasive paper) had to be carried out outside in the cold and sometimes wet weather of late November/early December. Once the rubbing down was complete, the locomotive was pulled into the paintshop, and the final painting work could begin.

The surface was again carefully cleaned with a solvent panel wipe to remove any oil and grease, and then one coat of a green oil-based paint was applied, followed by two coats of gloss – apple green and black on the relevant parts of the locomotive and tender.

When the initial painting was carried out at Darlington, final assembly was still going on, and it was possible to paint some of

the more difficult areas (e.g. wheels) before obstructions, such as the Motion, were added. Once in the NRM for final painting this luxury had disappeared, and access to the wheels in particular was tricky. The problem was compounded by the fact that the occasional drip of lubricating oil would land on the wheel surface – not an ideal companion for painting purposes. The problem was solved by cleaning each of the wheels immediately prior to paint application, for each of the five coats!

The next step was the detailed lining (black, white and red) and numbering/lettering. The lining was carried out by Ian Matthews and specialist sign writer Mike Thompson, while the numbering/lettering was added by Tony Filby. This marked a milestone for Tony, being his final project before retirement after 33 years as head of the painting and lining department at the NRM. The positioning for the lining and lettering was carefully measured out, and a chalk line was used to give a line to work to. However, the actual painting itself was carried out freehand – a job requiring considerable skill – and a very steady hand!

Each coat used around 10 litres of paint, and this was applied using 3in and 1½in brushes. To put the task into perspective, each coat of paint applied to one of the tender sides took approximately two hours. Many DIY enthusiasts will recognise the difficulty of painting a door at home, and maintaining a wet edge to the paint to allow the next brush application to flow smoothly in – just think about the challenge of doing this on a tender side!

The final stage in the process was to lightly rub down the surface with a 1200-grade wet and dry abrasive paper before applying two coats of a heat-tolerant varnish, which gives the locomotive its high-gloss finish, while also protecting the paintwork itself.

The finishing work was carried out against a tight deadline, with a covenantors' meeting arranged for 13 December 2008, at which a ceremonial unveiling was to take place – allowing those who had paid so diligently over the years to see the finished result.

The one item missing at this point were the nameplates – although these had long been made and displayed. An official naming ceremony was planned for the following February, and until then *Tornado* would run without a name, not unlike the locomotives of the original class when they were first introduced some 60 years earlier.

At the time of writing, the Trust has

announced that *Tornado* will be repainted over the winter maintenance period 2010–11 in BR 'Brunswick' dark green livery, hopefully producing a whole new set of photo opportunities for the enthusiasts of this livery.

Testing

Whilst most operational testing would be carried out during the commissioning period on the Great Central Railway and on the main line, obviously, considerable testing was carried out during the construction phases to ensure that the operating tests would be as trouble-free as possible. These included:

1. Boiler hydraulic and steaming tests. The hydraulic test involved taking the boiler above normal working pressure, while full of water, to check for leaks, and the steaming test (carried out in January 2008) involved lighting a fire in the boiler for the first time, and bringing it slowly up to steam pressure. This allowed a further inspection for leaks in a hot condition, and also enabled the safety valves to be tested.

2. Setting up of valve gear timing, which was carried out very successfully, by John Graham of NELPG (with whom the building at Darlington is shared – NELPG using one half of the building). The original set-up of the valve gear was assisted by a variable speed electrically driven set of rollers which rotated the coupled wheels with the engine lifted clear of the rails.

3. Testing of the various mechanical linkages including the reverser, steam regulator and cylinder drain cocks.

4. Testing of all the electrical circuits and components, checking for correct operation, and any excessive resistance in the cables, which indicates poor connections.

5. Extensive testing of all the brake systems including correct operation of the controls, and correct and free movement of all the mechanical linkages. The air brake cylinders themselves had already been tested at Meiningen prior to delivery.

This culminated in late July 2008, just days before the scheduled press launch, with the 'big test' as the locomotive moved under its own power for the first time. The configuration of the workshops and the specially laid 130 yards of track is such that the first move was in reverse gear.

After a short opening of the regulator and drain cocks to clear any water in the cylinders, followed by a short move and brake test, *Tornado* was taken the full length of the test track. There were varied expressions of relief and exhilaration from the engineering team, as the results of years of hard work finally came to fruition.

As the locomotive drew to a stop at the end of the track, and Director of Engineering David Elliott and his team celebrated the short but successful move, Operations Director Graeme Bunker (who was driving for the test runs) remarked: 'Well, I guess we had better try it going forward – it isn't going to make much money running in reverse!'

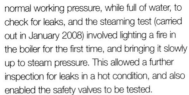

OVERLEAF

Resplendent in the new apple green livery, *Tornado* was posed on the turntable at the National Railway Museum at York in December 2008. *(David Elliott/A1SLT)*

60163

BRITISH RAILWAYS

RA 9

60/363   BUILT BY   1954
JOHN BOYD·CO·ENGRS·LTD
ANNAN   SCOTLAND

149

Chapter Four

Certification

The railway system in the UK has undergone major changes since the original Peppercorn A1s were operational. Track and signalling systems have been upgraded, electrification has been installed on much of the main line network, and main line expresses now run at speeds of up to 125mph, with greatly increased safety. The rolling stock itself has changed too, with modern diesel and electric traction producing a much improved ride for the travelling public.

OPPOSITE *Tornado* blows off at Sheffield during her York to Barrow Hill main line trial on 6 November 2008. *(Stephen Wright/A1SLT)*

From the outset, it was intended that Tornado should be built to standards that would allow it to operate over the national rail network, and clearly it was essential to get approval for this to happen. However, actual construction started in 1994, coincident with the privatisation of the railways, and a period of major change in regulations, areas of responsibility, and chains of command.

Prior to privatisation, the British Rail Director of Mechanical & Electrical Engineering (DM&EE), had approved the design of all new traction and rolling stock, to ensure that any new vehicle was compatible for operation on the railway infrastructure, and met rigorous standards for safety, both for employees and for the travelling public. The DM&EE had set these standards through a series of 'Engineering Instructions' that covered aspects of new design, construction and maintenance. Significantly, however, following the demise of scheduled steam locomotive workings in 1968, and after a short period when steam locomotives were banned from the network altogether, a selected number of preserved steam locomotives were then being accepted for working charter trains over the national rail system under what were known as 'grandfather rights', based on the historical fact that they had worked on the railway network for many years. The condition of these locomotives was regulated by a strict regime of six-monthly safety checks and stipulated periods for heavy mechanical and boiler overhauls, as laid down in later years in an inspection document, MT276. At privatisation of the railways this document remained in force and the grandfather rights continued.

When the rail industry was privatised in 1994, it was split into over 100 separate companies. Basically, the network infrastructure (track, signalling etc) was separated from train operations. The Health & Safety Executive (HSE) vested the responsibility for running a safe railway into the emerging company Railtrack. The engineering responsibilities of the former nationalised industry were broken up into a number of private companies dealing with various aspects of design, installation and maintenance, and within a number of these privatised companies were established

Conformance Certification Bodies (CCBs) and Vehicle Acceptance Bodies (VABs) which were then authorised by Railtrack to inspect and approve vehicles for running on the railway. In the first instance, their acceptance standards were based on the former engineering instructions, but as time progressed these were translated and rewritten into the Railway Group Standards, with an overriding set of instructions on certification procedures.

The CCBs and VABs were essentially an extension of Railtrack, to ensure vehicles complied with the standards. Each of the individual members of the CCBs and VABs were specifically authorised by Railtrack to issue a range of certificates, according to their experience and qualifications, and thereafter each individual was subject to regular audits by Railtrack, to ensure compliance with procedures and continued competence. To achieve 'Engineering Acceptance' to enable any vehicle to run on the railway system, the CCB was required to issue certificates for:

■ Acceptance of the design.
■ Acceptance that the vehicle had been constructed to the design standards.
■ Acceptance that the vehicle had in place a suitable maintenance scheme.
■ Acceptance that the vehicle was being maintained in a safe condition (as time elapsed and the system evolved, this latter requirement was incorporated into the work of the VAB).

On the satisfactory acceptance of these four certificates, the VAB was then authorised to issue a Certificate of Engineering Acceptance.

During the privatisation process a former division of the British Railways Board, Transmark, was bought by the Halcrow Group. Transmark had traditionally employed retired British Rail staff, mainly on a part-time basis, for contractual work, giving advice to foreign railway administrations, and they readily accepted the offer by four retired BR employees to form a CCB/VAB, essentially for steam locomotives, within their company. Each of the four had, at some time in their previous career within British Railways, been associated with the acceptance of preserved steam locomotives for running over the railway network, and so were not newcomers to the work involved. The group named themselves Heritage Engineering, and became authorised by Railtrack in 1995, for carrying out the whole range of certification duties for both steam locomotives and coaching stock. They quickly established a name for professionalism amongst the preserved steam railway fraternity, and despite competition from another similar organisation, cornered a major portion of the market.

With its steam pedigree this was a natural choice to become the CCB/VAB for *Tornado,* with the initial emphasis being on achieving certification for its design and construction. The lead contact for the A1 Trust within Heritage Engineering was Tony Broughton – a career railwayman, who had coincidentally served his apprenticeship at Darlington.

Tornado was to be a new-build locomotive of a type that had not run on the network for considerable number of years, and although it was to be substantially to the original design, because of the lapse of time, grandfather rights could not be automatically claimed. It was therefore recognised at a very early stage that a close consultation would be required with the relevant authorities throughout the build process, to ensure that no problems would be experienced when the locomotive was ready to run, on what was a unique and pioneering project destined to take several years before completion.

As the original Peppercorn A1 drawings were being used as the basis for the new build, it was useful at that stage to refer to the locomotive as a 'replica' and as such, it could be demonstrated that these locomotives had run safely and reliably during their short lives in the 1950s and '60s. Indeed, a paper to the Institution of Locomotive Engineers presented in 1961 by the President, J.F. Harrison, Chief Mechanical & Electrical Engineer for British Railways, contained the following paragraph:

'When one realises that these locomotives [the A1s] are better than the [LNER] A4 class, examples of which took part in the interchange trials in 1948, and which attained the best coal and water consumption figures per draw-bar horsepower/hour, one realises that these latter locomotives [the A1s] were

and in fact still are, perhaps the finest steam locomotives in the world.'

Early meetings took place with the relevant personnel in the Her Majesty's Railway Inspectorate (HMRI), which at that time had become a department of the HSE, and Railtrack, and it was established that there were no objections in principle, by either party, to the concept of building a brand-new steam locomotive for main line running. By using the word 'replica', a measure of agreement was also obtained with Railtrack that grandfather rights, as applied to other preserved steam locomotives, were applicable; an important step forward at that time, which meant that it was not necessary to produce or rework LNER design calculations.

It quickly became apparent, however, that a number of design changes were going to be necessary. The first had been undertaken very early on in the construction process by the replacement of each of the lapped and riveted frame plates with a one-piece frame. The second major change also arose at an early stage, from a declaration by Railtrack that they would not accept the design of riveted tyres as fitted to the original locomotives, and that only either a double-snip tyre (as had been fitted to SR Bulleid locomotives), or the Gibson ring fixing arrangement would be acceptable. In the event, the double-snip arrangement was accepted by the Trust, as described earlier, under Wheels, axles and tyres. It was also mutually agreed that modern air brakes would be required.

As time progressed, it became apparent that there were also a number of areas where the drawings were either missing or specified obsolete grades of steel or components, with notable examples being the Timken bearings which were originally oil-lubricated (now grease-lubricated), and various references to 'Best Yorkshire Iron'. Where the drawings were missing, the Trust, through David Elliott, was required to produce a new drawing from scratch. All were treated as design changes.

Similarly, changes in construction methods threw up a number of variations from the original build. For each of the major castings the wooden patterns had long been destroyed. Although most of the new patterns were made in wood, to have done this for all the small castings would have taken many hours and would have been prohibitively expensive. It was realised that where only one casting of each type was required a simple pattern of polystyrene could suffice, and used as in the 'lost wax' casting process. There were other similar changes with regard to cold riveting and the alternative use of fitted bolts. As the build progressed it was inevitable that some discrepancies in construction would also arise, which would require remedial action.

Any deviation from the original designs or non-conformance in construction required approval by the CCB, who in turn required the Trust to produce a reasoned risk assessment to prove that the change would not adversely affect safety. Meetings to review the changes were set up between David Elliott and Tony Broughton on a regular basis. This required the setting up of two registers to monitor progress towards acceptance, which were maintained and updated throughout the build process, the first one to cover design changes, and the second to cover construction non-conformances. Each of these registers eventually ran to many tens of entries, and formed key elements in the overall approval process. Any non-conformances had to be either signed-off as still acceptable by the VAB, had to be modified to achieve the acceptance of the VAB, or had to be reworked completely.

By the late 1990s, Halcrow had fully absorbed Transmark, and the latter name disappeared. Halcrow's attitude towards the use of retired, part-time engineers changed, with a move towards staff in full-time employment, and Heritage Engineering became incompatible with the new modern image that the company wished to create. By agreement of all parties, and subject to re-authorisation by Railtrack, Heritage Engineering, was transferred as a unit into The Engineering Link, a company formed out of the DM&EE during privatisation.

By agreement, Heritage Engineering was permitted to retain all its records and procedures and the move was undertaken seamlessly, without any disruptions for their clientele, or delay to on-going certification work. Relations with the Trust continued unchanged until 2002, when the four individuals within

Heritage Engineering took retirement, having first ensured that others within The Engineering Link had acquired the necessary authorisations from Railtrack to be able to continue the certification process. At this point, Tony Broughton, who had remained the key link to the Trust throughout, now surrendered his role to Bob Bramson. Tony Broughton then joined the Technical Advisory Panel within the Trust, to continue to advise on certification issues.

The Engineering Link continued to be the CCB/VAB for the Trust, albeit that it was taken over in 2002 by AEA Technology, and was subsequently sold on to a private equity group in 2006, when its name changed to DeltaRail. As DeltaRail, it saw the completion and certification of *Tornado*, and continues to be the Trust's VAB.

Around 2002, Graham Nicholas, a professional railway engineer and long-time supporter of the project, and employed by Railway Approvals (another VAB created after privatisation, and then part of EWS Ltd), volunteered his services to help with quality control and certification. The offer was gratefully accepted, and in 2004, Graham joined the Board as Quality and Certification Director.

Under Graham's guidance, a documented quality system, equivalent to ISO 9000, was established within the construction area. The discipline that this brought about in terms of working practices, and also in documentation, proved invaluable in supporting the certification process. It also imposed a discipline of evaluating and monitoring other suppliers to the project. It was decided not to go to full certification of the system under ISO 9000, due to cost, and the fact that the Trust was not producing work for sale, when such accreditation would have been more important. However, the system was audited and approved by The Engineering Link/DeltaRail

About this time, work had started on sourcing a potential boiler supplier. There were no boiler specifications in existence from the old British Railways days, and no British standard specifications for the design of a locomotive boiler either (in fact, few BS specifications ever existed, much being within the established practices of each of the former railway companies). The infrastructure for the manufacture of riveted boilers had long since been scrapped, and in addition, the arsenical copper for the inner firebox was becoming very difficult to obtain.

A further requirement of Railtrack was that the overall height of the locomotive should be reduced from 13ft 1in to 13ft 0in – a small change, but one which required a redesign of the safety valve plinth and the dome on the boiler. A combination of these factors led to the conclusion that a design change, and a move to an all-welded construction were needed. A side benefit of the move to a welded construction would give a reduction in weight, and this would assist in another area of concern. It was always suspected, although never formally declared, that the original Peppercorn roller-bearing A1s were actually overweight. As the weight of Tornado was going to increase further by the addition of an air compressor and other modern equipment, any reduction in weight elsewhere would be of considerable benefit. Subsequent weight measurements on Tornado have shown it to be within requirements.

An extensive period of researching potential suppliers, both in the UK and in Europe, had resulted in the identification of Dampflokwerk Meiningen in the former East Germany as offering the best option, as described under the Boiler section. The company is part of the Deutsche Bahn (German State Railways), and is one of 15 major works within the DB Engineering Group, which also carry out all major work on the present railway system, including maintenance of the ultra-modern high-speed ICE trains. This gave the company access to all the skills, advanced working practices, economies of scale, and quality assurance systems of a large, modern manufacturing operation. Several visits were made to Meiningen over the period 2002 to 2005, prior to final selection, including a full supplier audit carried out in 2005, which demonstrated to the full satisfaction of the Regulatory Authorities that they were able to design, construct and deliver a boiler to the required specification, and in full compliance with all EU standards.

Whilst there were no Railway Group Standards that related either to the design or construction of a steam locomotive boiler, it was listed as being a safety critical item, and as such it was destined to receive critical attention by all authorities. Full agreement by all parties was, however, quickly obtained when assurance was given that the boiler was to be built to European standards, and on completion would receive 'CE' plating from the Notified Body (NoBo) at Thuringen. Another hurdle was overcome.

One complication which arose at this time was the possibility of oil-firing the locomotive. Although none of the original Peppercorn A1s had used oil firing, other British locomotives had been run experimentally on oil, and this was common practice in other parts of the world, notably in parts of Europe, and in the USA and Canada, during the steam locomotive era. The oil used in most of these cases had been a residual grade (heavy oil) which requires preheating (both in storage and at the point of use), and an additional atomising medium at the point of injection into the boiler (either compressed air or steam). Whilst this was not a major problem during the steam locomotive era when locomotives were in operation for long periods without being cooled down, and the infrastructure in and around engine sheds was supportive, in the preservation era where engines are used intermittently, the potential problems were considered too great.

Further consideration was given to a relatively newly developed system for gas oil (diesel) firing, developed by Roger Waller in Switzerland. Whilst easier to handle than heavy oil, and readily available at all modern traction depots, this offered a different set of challenges, one of which was that it was relatively unproven (it had only been fitted to one main line steam locomotive, in preservation, in Switzerland and had only run a limited mileage in service). In addition, there were potential safety hazards in the event of an oil leak around the hot environment of an engine cab. This would have required extensive costs, not least in the certification requirements to prove that it could be safely implemented and managed, and although the option to add this at a later date was retained, it was rejected as a primary option.

In late 2002, following the Hatfield disaster and other aspects of poor performance, Railtrack as a listed plc went into administration. It was replaced by Network Rail, a private company, partially funded from revenues from the Train Operating Companies, and partially from Government subsidies, with all profits being ploughed back into the railways. Network Rail is now regulated by the Office of Rail Regulation (ORR), formerly part of Her Majesty's Rail Inspectorate (HMRI).

At about the same time, the Rail Safety & Standards Board (RSSB) was formed as a non-profit making organisation whose members comprised the Train Operating Companies, along with all other rail-related bodies. RSSB took over the responsibility for the maintenance and development of the Railway Group Standards, which among other things, laid down the standards to be met for all new vehicles running on the railway. The pains of the long planning and build process were now becoming all too obvious as the regulations were changed again. Fortunately, those parts which had already been approved retained such approval.

It was recognised by RSSB that steam locomotives would not be capable of meeting many of the requirements of the Railway Group Standards. Nevertheless, all 'grandfather rights' were withdrawn and a procedure was introduced whereby all non-compliances with the group standards had to be submitted for consideration by the RSSB's Rolling Stock Standards Committee, together with a statement of the safety risks involved and the measures to be taken in order that these risks were minimised. If accepted by the committee, a Certificate of Derogation would then be issued. This entailed a lengthy process of going through each of more than 40 standards, clause by clause.

The Railway Group Standards were of course, principally aimed at the design and construction and maintenance of fleets of modern diesel and electric locomotives and so contained standards for locomotives with very much smaller wheel diameters, clean cab environments, and a range of safety considerations never contemplated for steam locomotives. Notably, many standards involved design calculations relating to modern requirements, which included for example, such aspects as crash worthiness, noise emissions and track forces. Many such aspects had never been considered in the 1940s when the original Peppercorn A1s were being designed, but even

where they had been considered, records and calculations had long been destroyed. In these instances, a declaration was required within the application for derogation, to highlight that it was unknown whether the locomotive was compliant, or otherwise, as no data existed, but that the locomotive would nevertheless be safe to operate (often based on evidence of the safe operation of the original Peppercorn A1s).

The application for derogation for *Tornado* was submitted in 2004, and duly considered by RSSB, with Trust representatives in attendance, at which time each identified non-compliance with the standard was critically examined. The Trust representatives were required to justify the reasons for each derogation requests under fairly hard questioning. In many cases, the derogation was granted, but in others it was 'back to the drawing board' to reassess the position. As an example, the standards required that all traction units have a yellow visibility panel on the cab front (giving better visibility to anyone working on the track). It was agreed that this would be inappropriate for a steam locomotive and hence the request for derogation was granted. In another instance, however, it is stipulated that a headlight to a specified degree of brightness should be fitted – the RSSB would not accept any compromise in this instance, and derogation was refused. As a result, *Tornado* had to be fitted with the headlight, which is now a familiar sight.

The Certificate of Derogation was eventually awarded, allowing the certification process to continue, and permitting the locomotive to operate up to a maximum speed of 75mph.

Yet another complication was to occur in 2006. The European Union had brought out a series of directives on railway interoperability which were intended to harmonise the supply and operation of rolling stock within the different member states. These were translated into UK legislation with The Railways (Interoperability) Regulations 2006. This introduced a new certification body – the Notified Body (NoBo). During this period of significant change in the industry, discussions continued with the various authorities, with the final decision that *Tornado* must be treated as any other new railway vehicle, albeit that due note would be taken of its historical links. This meant that the new Interoperability Regulations would apply to the locomotive. Fortunately, DeltaRail were designated as a NoBo and could therefore perform the dual roles of VAB and NoBo.

By now, progress on the build was moving quickly, and January 2008 saw the successful live steam boiler pressure test. While a multitude of tasks were being carried out to complete the build ready for the first moves in steam, a detailed Testing and Commissioning Plan was being put together. This detailed the tasks to be carried out, in a progressive sequence, an approximate timing for the programme (obviously subject to change as the work evolved), who would be responsible for each task, and supplementary notes and explanations. This would provide further proof to the Regulatory Authorities of the professionalism of the team involved, and assist in gaining full accreditation for main line running.

The plan started with tests to be carried out on various systems prior to running (e.g. for correct operation of the air braking, tests for steam leaks under boiler pressure, etc). This was to be followed by low-speed running (25mph) light engine, and then low-speed loaded, gradually increasing to higher speed (60mph) tests, all at the Great Central Railway (GCR). The locomotive was then to be moved to the main

ABOVE With the Chairman, Mark Allatt, looking on, the Trust's Honorary Vice President, Malcolm Crawley, shovels the first coal into the firebox, preparatory to the lighting of *Tornado*'s first fire. *(Robin Jones/A1SLT)*

LEFT With Trust Chairman Mark Allatt looking on, Honorary President Dorothy Mather enjoys the first fire in *Tornado*'s firebox at Darlington in January 2008. *(Kathleen Eltis/A1SLT)*

line for a final series of tests up to 75mph. At each stage of the testing programme, detailed inspections would be carried out to ensure that no adverse effects were evident on the various components of the locomotive.

July brought a flurry of activity as the plan was put into action and *Tornado* was prepared for her first movement in steam on 1 August, watched by the world's press and many of the volunteers, contractors and suppliers who had been involved in the project.

Following a successful first operation, albeit on a very short length of track installed outside the workshop specifically for the purpose, *Tornado* was granted an official letter of authorisation from the HMRI to proceed to the next part of the testing plan, which was to take place on the GCR.

The first tests were of the braking systems, initially on a single inspection coach, but then on a rake of seven coaches. Loading tests followed, as the number of coaches was increased to 11, plus an idling Class 45 'Peak' diesel locomotive. All continued to go well and in mid-September, HMRI visited again to witness the locomotive in full railway operation. They were very satisfied with the results and issued a letter of approval for *Tornado* to enter passenger service on heritage railways. A few days later the locomotive hauled its

ABOVE The safety valves lift during the boiler steam test in January 2008. Graeme Bunker operates the regulator, with Barry Wilson and Malcolm Crawley looking on. *(Rob Morland/A1SLT)*

RIGHT David Elliott proudly holds the certificate alongside boiler inspector John Glaze, as the team celebrate the successful boiler test in January 2008. *(Rob Morland/A1SLT)*

ABOVE The A1 on the A1. *Tornado* was transferred from Darlington to Quorn & Woodhouse on the Great Central Railway by road, ready to start trials in August 2008. *(Alexa Stott/A1SLT)*

LEFT *Tornado* at Loughborough on the GCR having been towed from Quorn & Woodhouse. *(Neil Whitaker/A1SLT)*

first passenger train on the GCR. However, certification to run on the main line still had a few more hurdles to overcome.

The next very important tests were to demonstrate the track forces imposed by the locomotive when moving at speed, particularly on curves, when the forces would tend to cause a lateral movement of the track. DeltaRail, through its specialist vehicle dynamics department, fitted the locomotive with movement-measuring equipment wired back to an inspection saloon vehicle. Additional measuring equipment in the form of strain gauges was fitted to the sleepers on a chosen section of the track. *Tornado* then underwent a day's testing over this section of the track, with speeds being accelerated in stages up to 60mph for the first time. The electronic data collected was then computer analysed.

In order to provide comparison data, the tests were repeated a few weeks later with 'Britannia' 4-6-2 No. 70013 *Oliver Cromwell*, a

preserved locomotive already certified for main line running. The comparisons showed that No.60163 was superior in almost all respects, and thus another milestone was passed with flying colours.

Further tests were carried out on the AWS/TPWS warning systems, and also on the air-braking system. Again, all tests were passed with no problems at all.

The next stage of the testing had to take place on the main line, but before this could happen, paperwork had to be received, giving the locomotive temporary authorisation to run on the main line for the tests to be carried out.

In addition, further minor items of work had to be carried out on the locomotive before it would be allowed to run on the main line. These included the fitting of a spark arrestor (to reduce the chance of starting lineside fires), and the fitting of overhead electric line warning notices. This work was carried out at the NRM in York.

At the end of October, the first Engineering

LEFT The track was instrumented, ready for *Tornado*'s test operations at the GCR, September 2008. *(Owen Evans/A1SLT)*

LEFT The Trust's Graham Nicholas in discussions during testing operations at the GCR, September 2008. *(Owen Evans/ A1SLT)*

locomotives with the same last two digits. The main line tests could now start.

There were three test runs to be carried out, a light engine run, a 60mph loaded run, and a high-speed loaded run, in this case up to 75mph. These obviously had to be fitted around timetabled use of the network, when suitable paths were available, and to accommodate this, the runs were scheduled in the late evening and into the night. The light engine run was from York to Scarborough and back, closely followed by the first loaded run on a train comprising 12 Mark 2 coaches and a supporting idling diesel locomotive (a load in excess of 500 tons). This time, the run was from York to Barrow Hill (Chesterfield) via Sheffield. The run itself was text book, and on the way back things were going so well that the speed was taken up to the 75mph. However, on closer inspection of the locomotive following the run, it was found that the white metal in the crosshead bearing of the inside motion had overheated and run.

Subsequent analysis, and a check back in the history books, revealed that this was not an uncommon occurrence on all the LNER three-cylinder Pacifics, and modifications had been carried out in service to improve the lubrication to this bearing. Similar modifications were quickly put in place on *Tornado*, and less than two weeks later the locomotive was ready for its third test run. Not surprisingly, the operator for the tests (EWS) requested a light engine run first, to assess the results of the modifications to the inside motion. This was carried out successfully a few hours before the scheduled 'big' test run on 18 November.

For this run, DeltaRail had re-instrumented the locomotive to allow further measurements to be taken on the 75mph, high-speed run. The train this time consisted of ten coaches plus the supporting idling diesel, and was from York to Newcastle and return. Once again, the run itself was flawless and impressive, although on examination the following day it was found that there had been another white metal failure, this time on an outside crosshead. This was not down to lubrication failure however, but was due simply to insufficient working clearances. This was quickly rectified.

Part of the run involved passing over a

ABOVE Monitoring equipment was set up in a saloon coach for the GCR testing. *(Owen Evans/A1SLT)*

Acceptance Certificate was signed, and this was quickly followed by the Network Rail Route Acceptance Certificate. *Tornado* was entered into the TOPS (Total Operations Processing System) database, essential before any vehicle can run on the main line network, as number 98863. The first two digits, 98, identify the vehicle as a steam locomotive, the third digit, 8, its power classification, and the final two digits, 63, identify the individual locomotive. When possible, the last two digits of its painted running number (60163), although this is not always possible due to duplications of

Wheelchex ® site south of Thirsk, which measures instantaneous forces on the track as the train passes over. Concern had been expressed for some time about the magnitude of vertical 'hammer-blow' forces from steam locomotives which had been blamed as a cause of track problems for many years, but little attempt had been made to measure the forces imposed. What was known was that three-cylinder engines were generally better than two-cylinder engines, because of their more evenly distributed crank centres at 120°. These forces had been computer modelled during the construction process, and particular care had been taken with the static balancing of the wheelsets, thus the Trust was therefore hopeful that the tests would show favourable results. In practice, readings from the site showed that far from the locomotive presenting a problem, the highest peak forces were actually recorded by the other vehicles in the train, not the locomotive itself!

Although the test runs had been carried out at some rather anti-social hours, the lineside interest generated by these was incredible – crowds had been present at stations, on

ABOVE No. 60163 *Tornado* arrives at Scarborough after its first York–Scarborough main line test run on 4 November 2008 *(D Bailey/A1SLT)*

RIGHT Despite the late hour, crowds lined the platforms at Durham as *Tornado* stormed through on the 75mph test run to Newcastle on 18 November 2008. *(Andy Graves/A1SLT)*

BELOW *Tornado* stands in Newcastle station following the test run from York on 18 November 2008. On the footplate are driver David Court and the Trust's Graeme Bunker. *(Andy Graves/ A1SLT)*

bridges and other vantage points to see *Tornado* for the first time on the main line.

With the tests now complete, certification was in the 'home straight', and it was a matter of completing various administrative tasks to get the official documents in place ahead of the first scheduled, main line passenger-hauling run on 31 January 2009.

The maintenance plans were completed, final items from the 'snagging list' were fixed, and all the paperwork from the various tests and certifications was assembled into a NoBo technical file, required under the Interoperability Regulations. Finally, DeltaRail signed off the paperwork on 22 January 2009. The next day, DeltaRail issued full Engineering Acceptance, followed by Network Rail Route Acceptance certification on the 26th. On the same day, the Trust issued its own Declaration of Verification (an essential document required under the Regulations). Finally, on the 27th, the ORR issued the Authorisation Letter allowing *Tornado* to enter into service. A massive milestone had been reached, more than 18 years after the adventure had started.

However, this did not quite complete the certification story. Currently, there is a blanket regulation that all large-wheeled steam locomotives (over 6ft 2in diameter) on the national network have a speed limit of 75mph. Smaller-wheeled steam locomotives have lower maximum permissible speeds. This is covered by the Railway Group Standard GO/RT3440, Steam Locomotive Operation, which deals with those unique aspects of operating steam locomotives, which are not covered by other Group Standards.

GO/RT3440 is in the process of being reviewed and updated, and the Trust is actively involved, along with other steam locomotive operators, in trying to get the speed limit raised under appropriate circumstances.

Tornado was designed to run up to 90mph, and at the time of writing there is still hope that this will happen in the not-too-distant future.

OFFICE OF **RAIL REGULATION**

Mark Allatt
Chairman
A1 Steam Locomotive Trust
Darlington Locomotive Works
Hopetown Lane
Darlington
DL3 6RG

Your ref: -
Our Ref: Case: 4147726
File ref: RI/58/1/108/1996-03
EIN/NVR UK/51/2009/0001

Date: 27 January 2009

Mike Holmes
HM Principal Inspector of Railways
HMRI
ORR
The Pithay
Bristol
BS1 2ND
Direct Line: 0845 301 3566
e-mail: mike.holmes@orr.gsi.gov.uk

Dear Mr Allatt

THE RAILWAYS (INTEROPERABILITY) REGULATIONS 2006

Authorisation of A1 steam locomotive 60163 'Tornado'

I refer to your application for authorisation, received under cover of the letter dated 26 January 2009.

Following review of your application, I can confirm that ORR grants authorisation under Regulation 4(1)(a) of the above Regulations for the placing into service of the A1 steam locomotive 60163 'Tornado'.

The conditions and restrictions of operation are those stated on the declaration of verification dated 26 January 2009 and contained in your technical file A1 Tornado, LD78051_A1_TF Issue A Rev 0, submitted on 26 January 2009.

The rolling stock authorised by this letter must be operated and maintained in accordance with the essential requirements detailed in Regulation 12(1) and 12(2) of the above Regulations. In addition, the technical file must be kept up to date and the vehicle details placed on the appropriate registers. Any future modifications to the rolling stock may be considered to be a renewal or an upgrade. Any major renewal or upgrading work must be authorised by the ORR prior to bringing into service.

Head Office: One Kemble Street, London WC2B 4AN T: 020 7282 2000 F: 020 7282 2040 www.rail-reg.gov.uk

---

As the Contracting Entity, I remind you of the need for you to have adequate arrangements within your safety management system to control the risks associated with the introduction and handover of this rolling stock.

Yours sincerely

Stan Hart
Head of Inspection - Engineering
HM Railway Inspectorate

Copies: Mr C Carr, Deputy Director, Rail Standards & Safety, DfT, Zone 32, 4th Floor, Great Minster House,
76 Marsham Street London SW1P 4DR
Mr P. Hooper, Interoperability & Standards Manager, ORR, One Kemble Street, London WC2B 4AN
Mr M. Holmes, Principal Inspector RVNET, HM Railway Inspectorate
Mr D Keay, LRT and Minor Railway NET Manager, HM Railway Inspectorate, ORR Birmingham
Mr G Nicholas, A1 Trust

Chapter Five

Maintenance

In common with every other steam locomotive, *Tornado* is a complex piece of machinery, with many moving parts. By virtue of what it is, it also has a number of safety critical systems, most noticeably the firebox and boiler, which operate at high temperature and pressure, and represent a significant potential safety risk to both crews and the general public if anything goes wrong.

OPPOSITE Trust engineer Peter Neesam oils up the motion, while in the background, coal is being loaded into the tender. *(Rob Morland/A1SLT)*

Add to this the fact that the locomotive operates on the public railway network, at high speed, and it is obvious how critical correct maintenance becomes. Even failures of the locomotive in service which do not have a safety implication, could lead to serious and unacceptable delays for service trains if the locomotive is blocking the line.

Although accidents did happen in the steam locomotive era, compared to the number of locomotives in operation, and the number of passengers carried, these were rare. Lessons were learnt and improvements implemented as a result of each incident (both in the steam locomotive era and subsequently), and have led to rail travel becoming the safest form of public transport. To ensure that this level of safety is maintained a key element of allowing a steam locomotive to run on the main line is that the operator has, and adheres to, a comprehensive and effective maintenance plan for the locomotive.

In the operating era of the old Peppercorn A1s, there were numerous 'sheds', where locomotives were maintained and prepared for operations, including watering, coaling, de-ashing, cleaning, and lubrication. These sheds had staff skilled in various essential disciplines to carry out other minor repairs, and the day-to-day maintenance requirements of steam locomotives was a routine operation. Maintenance staff soon became familiar with the peculiarities of individual classes of locomotive (and in many incidences the peculiarities of individual locomotives themselves). These were backed by the 'Works', where locomotives were built and major overhauls were carried out, the largest on the LNER being Doncaster. Even at this time, detailed records were kept on cards of all significant work carried out on a locomotive.

With the demise of steam, these facilities were lost, along with most of the experienced staff, and the heritage movement has been faced with gradually rebuilding an infrastructure

BELOW *Tornado* in the workshops at the National Railway Museum at York during it's winter maintenance period January 2010 *(David Elliott/A1SLT)*

in terms of facilities and staff to ensure that the rigorous safety requirements of the Regulatory Authorities are met.

The operation of steam locomotives on the main line in the preservation era has always been supported by an inspection regime from the railway authorities. In the early days of steam preservation, this was primarily a continuation of the checks and examinations which were carried out during the BR steam era, usually by the same inspectors.

However, there was considerable regional variation and so, in 1984, these inspections were standardised into a document known as MT276 'Examination Schedule for Preserved Steam Locomotives Running on the Main Line'. This was revised again in 1990, and has remained the 'bible' for main line steam locomotive inspection standards to the present day (surviving, and indeed being enshrined in, the post-privatisation world via reference in Railway Group Standard GM/RT2003

'Certification Requirements for Registration of Steam Locomotives' [1996]).

The Trust was very aware of this background when considering the compilation of a suitable maintenance scheme for *Tornado*. However, MT276 has its limitations, having not been revised for nearly 20 years, in consequence of which it did not include reference to maintenance of the more modern operational safety equipment now fitted to main line steam locomotives, such as the air-braking system, TPWS and the OTMR data recorder. In common with the approach elsewhere with the locomotive, the Trust was keen to adopt contemporary techniques where possible and therefore the maintenance plan that was created was a blend of old and new. Its layout and structure follows that of a modern locomotive – examples studied in its development were Class 66 diesel-electric and Class 90 25Kv ac electric. However, the content closely followed the wording and guidance

BELOW The front bogie is hoisted out of the wheel-drop at the NRM during winter maintenance January 2010 *(David Elliott/A1SLT)*

of MT276, supplemented by the additional instructions for the modern equipment.

The majority of the work was undertaken during 2008 and involved many hours of painstaking voluntary effort by covenantor Joe Brown (who compiles such documents for a living), overseen by Graham Nicholas. Additionally, Tony Broughton (who had been involved with the development of MT276 some 20 years earlier) provided valuable support and guidance, not least of which was supplying a copy of the original BR maintenance specification for steam locomotives, MP11.

The resulting comprehensive maintenance plan for *Tornado* was a key part of the certification process, and further demonstrated the professionalism of the Trust to the Regulatory Authorities. The schedule is made up of a series of documents, running to several hundred pages. Starting at the highest level, the plan details the areas which require inspection/action and the frequency with which each is required, working down to detailed procedures for each check or maintenance action. This is supplemented by check sheets which must be completed and signed-off each time any inspection or maintenance task is carried out, to provide a documented record available for inspection by the relevant authorities at any time.

The requirements are basically split into six schedules, driven by the frequency with which each must be carried out, very similar to (but obviously more extensive than) those on a car. These are:

1. Daily 'Pre Run' or 'S' examination – to be carried out before every run.
2. Safety Examination 'A' – to be carried out every 14 operating days.
3. Enhanced Safety Examination 'B' – to be carried out every fourth Safety examination.
4. Annual Examination 'C' – to be carried out every 12 months.
5. Intermediate overhaul – to be carried out every five years.
6. Full overhaul – to be carried out every ten years.

The daily S exam has a total of 32 items which must be checked and signed off before the locomotive is allowed to run. Many of these are visual checks of the locomotive and tender, which would look for issues such as excessive wear, loose bolts, any signs of fractures appearing, controls and instruments sticking, or not functioning correctly, etc. Many of these would be attended to and corrected on the spot (e.g. minor steam or water leaks), but the identification of a potentially more serious issue, which cannot be fixed on the spot, could well lead to the locomotive being failed and not being allowed to run until the problem is fixed. Part of the daily tasks include checking oil levels in automatic lubrication systems, manually lubricating where required, and topping-up the sanders. It also covers water quality tests on the boiler feedwater.

The A Safety examination has a total of 24 checks and tests which must be carried out, some of these with the engine in steam, and some with it out of steam and cold. As opposed to the visual inspection of the daily test, this focuses much more on physical testing of the operation of various systems on the locomotive. For example, the cylinder drain cocks, injectors, and brake tests. It also covers a boiler washout and an internal firebox examination, along with the cleaning out of water filters on the water feed system, and a test on the boiler safety valves (that they lift at the correct pressure and can pass the required amount of steam to prevent boiler pressure building any further, and that they close off fully).

The B Safety examination covers all the items in the A examination, plus a further 20 points, some of which must again be carried out with the locomotive in steam and some with it cold. Additional items include measurements (and recording) of the height at various points around the locomotive. Variations from normal (either of a single-point reading or of several readings) may be an early indication of a problem with the springs. This examination also involves a degree of dismantling to facilitate cleaning and internal wear checks.

The C Safety examination carried out every 12 months, includes all the items in the B examination, plus a further eight points. Three of the key additions are a piston and valve examination (opening of the cylinders and valve chests to inspect for wear), an inspection of

the connecting rod bearings, and an internal inspection of the tender water tank. This examination is carried out over the winter maintenance period when the locomotive is not in service.

The five-year intermediate overhaul is more extensive and involves a greater degree of dismantling. Detailed inspections will be carried out on all major components, looking for signs of excessive wear, or stress fractures. This will involve some ultrasonic testing to ensure there is no hidden deterioration. It was anticipated that this will involve the replacement of some boiler stays, and the likely replacement of at least some of the small boiler tubes.

The ten-year major overhaul is what was known as a 'heavy overhaul' in the steam locomotive era. This will involve an extensive strip down, including the removal of the boiler.

The boiler itself will undergo extensive work, including the replacement of all small tubes and most stays, and ultrasonic testing and repairs as necessary of the shell. All other major components on the locomotive will be removed, checked and repaired as required, before it is rebuilt and tested in an 'as-new' condition. At this point, it will be awarded a further ten-year boiler certificate and return to service.

Careful thought during the construction of *Tornado* has simplified some maintenance tasks compared to the original Peppercorn A1s, and early operation has identified areas where further improvements can be (and in some instances, have been) made. A number of these will be made in the future. However, the locomotive is a complex piece of machinery and close attention to maintenance will always be needed to keep it operating safely and at peak efficiency.

ABOVE The coupling rods are hoisted back into position during winter maintenance January 2010. *(David Elliott/A1SLT)*

Chapter Six

Organising and operating a main line rail tour

From the early days, it was always planned that *Tornado* would be a working engine, partially to generate revenue to offset the cost of the build and the subsequent maintenance of the locomotive, but primarily to give the public the opportunity to see, and ride behind, one of these iconic express locomotives, travelling at speed.

OPPOSITE Support crew member Steve Wood polishes the top of the boiler. *(Rob Morland/A1SLT)*

125

The A1 Steam Locomotive Trust was primarily established to build and operate the locomotive, rather than to run rail tours – a market where there were already a number of well-established and respected operators, using a variety of preserved steam traction. It was therefore decided that most of the operating time available would be taken up by leasing the locomotive to haul specific excursions organised by these operators. In addition, many of the private heritage railway operators were keen to have *Tornado* for short periods of time as a visiting engine, hauling trains over their preserved lines.

In these circumstances, operating crews are organised by the Train Operating Company or the heritage railway involved, although there is always an 'Owner's representative' on the footplate to help look after the locomotive and provide advice to the crews.

However, the Trust did decide that it would run a small number of trains each year under its own auspices, and this chapter will look at the work involved behind the scenes to create that 'magic' day out for the travelling public. The procedures involved may vary slightly in detail with other tour operators, but the principles involved will be the same.

The initial step for any tour is to decide the start and end points. Several factors have to be taken into account in selecting an appropriate run. First, it has to be practical – can the distance be covered within a reasonable timescale, given the likely amount of service traffic on the line, and taking into account watering stops etc, and will it result in sufficient time at the destination point for the locomotive and train to be serviced and readied for the return journey. Secondly, and most importantly, is it likely to be commercially viable – organising and running an excursion is a very expensive exercise, and the Trust cannot afford to incur a financial loss on an excursion.

This is a balance between what the ticket prices need to be, and the likelihood of selling all the available seats. Generally, a high occupancy is needed to cross the line between running at a loss and running at a profit. On most tours a selection of seat prices is offered to attract the widest possible audience – Standard Class (seating at tables for four, with a buffet car service for refreshments), First Class (upgraded seats at tables for four, with a welcome drink and complimentary tea and coffee throughout the journey), and Premier Dining (upgraded seats at tables for two or four, with champagne breakfast and four-course dinner included, along with complimentary tea and coffee throughout the journey).

A key factor is of course the tour route and destination, with one or other (or preferably both), being of significant interest to attract the passengers. A further consideration, connected to these, is the number of potential pick-up points, as each stop will increase the journey time, but will increase the potential customer base.

The next step is to obtain quotations from a licensed operator (Train Operating Company) to run the tour. The operator will obtain the necessary approvals from Network Rail, and will provide the coach rake at the appropriate point, along with back-up diesel power where required, and the train crew(s) – driver, fireman, traction inspector (depending on the Train Operating Company concerned) – and guard. The two key operators for steam traction in the UK are DB Schenker (whose primary activity in the UK is rail freight), and West Coast Railways (set up specifically to run rail charters).

Once the quotations are received, a decision can be made as to whether the excursion will be viable, taking into account the other costs to be incurred directly by the Trust as the tour operator. Once the decision is taken to go ahead, then the organisation moves into the next phase – the key step now being to publicise the trip and sell tickets as soon as possible. Not being a regular tour operator, the Trust subcontracts the ticketing to one of the well-established rail tour companies, usually Steam Dreams. This facilitates the use of their web-based ticketing facilities. The Trust publicises the trip via its own website, and in all communications to its covenantors.

As the day approaches there are still a number of other tasks to undertake. The catering is subcontracted to one of a number of established catering companies. Quotations are obtained, menus are discussed and refined, quality is assessed, but this is not usually an issue as most of the catering companies have a well-established reputation, and the catering company is contracted.

Water and coal for the locomotive is the next issue. Coal will be required at the commencement point and at the turn-round point, and this is ordered for delivery by one of a list of approved coal merchants, depending on the destination. Water is slightly more complicated as most excursions will require at least one en-route stop for water top-up, some will require more. The first step is to consider mileages and appropriate stopping points, which must not block the main line for service trains. This therefore is usually an appropriate station with a bypass line, a servicing yard with entry and exit connections, or even a road overbridge, where a road tanker can position on the bridge above the tender. If a fire hydrant is available, this is the most convenient source of water, albeit more labour intensive to use. More usually, the water will be delivered by a road tanker with integral delivery pump. Again, there is a list of approved and trusted suppliers who are contracted to provide the service when required. One such company normally provides support to the fire service – topping-up fire engines at an emergency incident. The pumps available on their vehicles allow 5,000 gallons to be delivered into

Tornado's tender in six minutes, using the twin delivery pipes.

The next consideration is staffing. Although the train operator provides the qualified train crew (driver, fireman, traction inspector and guard), the locomotive owner has to provide the support crew, and the tour operator (in this case the same company – the Trust) has to provide the train stewards. In the case of the Trust these are all volunteers, usually covenantors, who give their time to support Tornado. The support crew will be responsible for looking after the locomotive throughout the day's run – the visual check at each stopping point, organising the watering at each watering stop, and the main checks and service at the turn-round point (lubrication, cleaning the fire, coaling and watering etc). Support crew members must hold a valid PTS (personal track safety) certificate, acquired by attending a certified training course.

At least one steward per coach must be provided. The primary responsibility of the stewards is to ensure the safety of the passengers – checking the doors are closed and the secondary locking (usually in the form of slide bolts) is in place before the train moves, ensuring passengers do not become over enthusiastic in terms of hanging out of

BELOW The support crew preparing for *Tornado*'s first public main line run on 31 January 2009. *(Rob Morland/A1SLT)*

windows for photographs (always an issue on steam-hauled trains), and that when arriving in a station, doors are not opened before the train has come to a stop. At the turn-round point, the stewards assist on the platform, helping passengers on and off the train, and seeing that they do not stand too close to the platform edge during train movements. However, the stewards also have a secondary role, that of helping passengers to enjoy the day out – answering questions (usually, in the case of *Tornado*, about the locomotive itself), and generally being of assistance when required.

Obviously, the main reason most people choose to travel on a Trust excursion is to be hauled by *Tornado* itself, and aside from the train, the work needed to run the locomotive on a day excursion is substantial, and starts well before the actual day.

Assuming the locomotive is cold prior to the trip, preparatory work will normally start three to four days before the actual day of the run. The Trust's duty engineer will start with the S examination, which involves a detailed examination of the locomotive, including the insides of the firebox and smokebox, completing a checklist sheet. Providing no problems are discovered (or at least none which cannot be fixed on the spot), then the boiler is prepared for starting the fire. The boiler water level is set, usually by draining part of the boiler contents, although after extended maintenance activity where the boiler has been drained it would consist of filling the boiler via the blowdown valve. Water quality is tested and where necessary, chemical additions are made to the tender to correct pH. A fire is then lit in the firebox to allow the boiler to warm up slowly over two to three days. This avoids excessive stresses in the boiler due to rapid warm up.

With a full day to go, the rostered support crew arrive – normally there would be up to eight on a trip including the responsible officer and the duty engineer. The crew set to work immediately as the locomotive has to be cleaned, oil and

sandboxes topped-up, the support coach has to be prepared, and food and drink provisions have to be obtained for the support crew for the duration of their 'on-duty' time. The tender has to be coaled, this usually being delivered on a lorry with a Hiab grab so that it can be off-loaded directly from the lorry to the tender. The tender also must be filled with water.

The day before the run, the train operator's fitness-to-run (FTR) examiner will arrive to carry out an in-steam inspection of the locomotive and support coach. This is a two to three-hour process during which a detailed examination will be carried out, including a thorough visual inspection, followed by a braking system test, testing of boiler safety valves, etc. It is vital that an inspection pit is available to allow the inspector to complete his work. At the end of the inspection, assuming no problems, then a 'Fitness-to-Run' certificate is issued, applicable to that specific run only. However, before the locomotive is allowed to start the run, a copy of this certificate must be in the hands of the Train Operating Company's locomotive engineer, so access to a fax machine is another important consideration.

And so the day arrives, and shortly before the designated movement time the train crew arrive, consisting of the driver, the fireman and the traction inspector. On occasions, if the driver is unfamiliar with the route, a fourth man – the pilotman (who would have route knowledge) – will join the crew.

With the crew on board, including the owner's representative (who will have detailed knowledge of the operation of the locomotive), it is time to move off and pick up the stock – often with the support of a diesel locomotive on the other end of the train to pull it into a terminus station. Statutory brake tests are carried out once the stock is coupled up, and clearance is then awaited to move the train into the platform.

With passengers on board the trip gets underway, and it is chance for the support crew to have a short rest before they are called on to assist at watering stops or any other stops – and particularly with anything unexpected which can crop up from time to time.

Once at the destination, the support crew are back in action. The stock is normally moved off to a carriage siding, and the locomotive

moved to a service point, where coaling and watering take place again, ready for the return run. The fire is cleaned, removing any clinker and ensuring a clean even fire bed. A further inspection is carried out around the locomotive to check for any problems, lubrication pots are topped-up, and sandboxes are checked and topped-up if needed.

It is then time to collect the stock and move into the station to pick-up for the return run.

Once back at the start point, and with the tired (and happy) passengers off the train and on their way home, the stock is taken back to the yard, and the locomotive and support coach then return to the prearranged depot. There is still work to be done to 'dispose' of the locomotive before the support crew can call it a day. The fire is left in the locomotive, but with the ashpan dampers closed, which allows the fire to die gradually, and the boiler to cool slowly – again, to limit mechanical stresses. The boiler is filled to the top with water, to compensate for the contraction of the water as it cools.

Although based on the original Peppercorn A1 design, a lot of thought was given to operations during the construction phase of *Tornado*, thus resulting in disposal at the end of a run being a comparatively easy exercise, particularly when compared with a number of other heritage locomotives. A very tired support crew are doubtless very grateful for this at the end of a long day.

ABOVE Blues and Two Tones' tankers supply the water during a watering stop at Grantham. *(Rob Morland/A1SLT)*

Chapter Seven

The first year in service

━━━◉━━━

The years of trials and tribulations all came to a pinnacle on 1 August 2008, the day *Tornado* made her first public moves on the short length of specially constructed track outside the Darlington workshops where construction had taken place.

OPPOSITE In the summer sunshine, *Tornado* passes Dawlish sea front with the 'Torbay Express'. *(Ian MacDonald/A1SLT)*

131

After short speeches by Mark Allatt, Chairman of the Trust, and Ian Haszeldine, the Mayor of Darlington, and with Honorary President Dorothy Mather (the widow of the locomotive's original designer Arthur Peppercorn) on the footplate, and in front of the assembled world's press, the drain cocks were opened (enveloping the assembled hoards in a cloud of steam), the regulator was notched open, and *Tornado* gently moved off. It was a truly emotional moment for thousands of steam railway enthusiasts the world over – the first main line steam locomotive to be built in the UK for almost 50 years.

After a couple of weeks of testing and adjustment at the works (including the completion of minor outstanding jobs), and during which time, first the covenantors and sponsors and then the general public, had the chance to witness the locomotive at close quarters, *Tornado* was carefully moved on to a low loader and away from its birthplace. It was soon a case of the 'A1 locomotive on the A1 road' as the transfer to the Great Central Railway at Loughborough was under way. With the precious cargo off-loaded on to GCR metals, testing soon commenced.

The next few months competed the testing, both at the Great Central, and subsequently

ABOVE Flanked by Trust Honorary President, Dorothy Mather, and the Mayor of Darlington, Ian Haszeldine, Trust Chairman Mark Allatt addresses the press and supporters prior to the first moves on 1 August 2008. *(Chris Milner/A1SLT)*

RIGHT With Dorothy Mather in the cab, No. 60163 *Tornado* makes its first move on 1 August. *(Chris Milner/A1SLT)*

RIGHT Outside Hopetown Works No. 60163 made numerous runs back and forth along the 130 yards of track during its 'first moves' weekend in August 2008. *(Neil Whitaker/A1SLT)*

on the main line, followed by the painting into the apple green livery in the paint shop of the National Railway Museum at York. Another emotional moment occurred on Saturday, 13 December 2008 when the covenantors' annual convention, held at York, culminated in an unveiling ceremony in the Great Hall of the NRM. *Tornado* was resplendent in her new livery, completed only hours before.

During this time, plans were being made for the first main line public passenger-hauling runs, taking place on 31 January 2009 (York to Newcastle and return), and 1 February (Doncaster to Durham and return). These trains, aptly named 'The Peppercorn Pioneer', were specifically organised to allow covenantors and sponsors the opportunity to experience *Tornado* on the main line for the first time. The locomotive performed immaculately, and the advance publicity generated by both the general and the steam-specific media captured the public imagination, and crowds turned out to witness the historic event. Station platforms were packed, and every bridge, road crossing and other vantage point saw ecstatic crowds wave the trains through.

ABOVE *Tornado* hauled its first public train at the Great Central Railway on 21 September 2008. *(Ken Richardson/A1SLT)*

LEFT The new apple green livery was unveiled to the covenantors at the National Railway Museum on 13 December 2008. *(Les Dixon/A1SLT)*

However – the 'big one' was to follow the next weekend when, on Saturday, 7 February *Tornado* was set to haul 'The Talisman' along the East Coast Main Line from Darlington to London, King's Cross. The intervening week had seen snow down the eastern side of the country, and many places along the route still had lying snow.

Although covenantors were given priority booking for this run, it was also the first main line run open to the general public. Leaving Darlington in the early morning, *Tornado* steamed southwards along the route which had been the stomping ground of the original Peppercorn A1s in the 1950s and '60s – York, Doncaster, Retford, Grantham, Peterborough

and through the London suburbs and into King's Cross itself – the first time the famous terminus had seen an A1 in some 45 years. Once again, crowds flocked to every vantage point en route (it was estimated that more than 30,000 people had turned out to see *Tornado* at some point on its run that day), and the platform at King's Cross was packed with around 3,000 people as the train drew to a halt a few yards from the buffers. What was particularly pleasing was that this was not just a crowd of older steam enthusiasts – this was the great British public in every sense of the word. The 'Tornado Effect' was well under way!

Following a number of charter runs around the South of England, *Tornado* returned to

ABOVE Pipers welcome No. 60163 at Newcastle station following its first public main line run on 31 January 2009. *(Chris Calver/A1SLT)*

LEFT Having taken on water in Peterborough Yard, No. 60163 approaches the station at the head of the 'The Talisman' on 7 February 2009. *(Steve Philpott/A1SLT)*

York in time for the next major event in her life. On 19 February, and in front of a specially invited audience at York railway station, the locomotive was officially named by Their Royal Highnesses, The Prince of Wales and The Duchess of Cornwall. A short speech by the Prince of Wales, periodically hidden by drifting clouds of smoke from the locomotive's chimney, preceded the unveiling of the nameplates. In a further tribute, the RAF provided a flypast of a Tornado F3 accompanied by two Hawk jets. The observant enthusiast will have noticed that there is a difference between the nameplates

ABOVE With Trust Chairman Mark Allatt and HRH Duchess of Cornwall looking on, HRH The Prince of Wales speaks at the official naming ceremony at York on 19 February 2009. *(Neil Whitaker/A1SLT)*

BELOW Speeches over, and the *Tornado* nameplate is officially unveiled by HRH The Prince of Wales. *(Neil Whitaker/A1SLT)*

RIGHT 'Now, which lever is it?' Following the naming ceremony, and in appropriate dress, HRH The Prince of Wales takes the driving seat at York station. *(Rob Morland/A1SLT)*

ABOVE Now officially
carrying nameplates,
Tornado's first duty
was to haul the
Royal Train from
York to Leeds on
19 February 2009,
pictured here passing
Church Stretton. *(Ian
MacDonald/A1SLT)*

on either side of the engine – each bears a
different RAF crest, one representing RAF
Leeming, and the other RAF Cottesmore – two
of the key Tornado aircraft bases. Now officially
carrying its nameplates for the first time,
Tornado's next duty was to haul the

Royal Train from York to Leeds – again a
faultless performance.

The latter part of February saw the
locomotive heading to the northern end of the
East Coast Main Line with a York to Edinburgh
charter, returning a week later on a further

RIGHT *Tornado* stands
with fellow LNER
Pacifics, A4s *Sir Nigel
Gresley* and *Union of
South Africa*, and A2
Blue Peter, at Barrow
Hill's steam gala in
March 2009. *(Clive
Hanley/A1SLT)*

working. After further main line charters at the beginning of March, *Tornado* was moved to Barrow Hill at Chesterfield for a steam gala which also featured the other East Coast Pacifics *Blue Peter* (A2), *Sir Nigel Gresley* (A4), and *Union of South Africa* (A4).

On into April, the highlight of which was the filming of the 'Race to the North' for the BBC TV programme *Top Gear*, featuring Jeremy Clarkson and which pitted *Tornado* against a Jaguar XK120 and a Vincent Black Shadow motorcycle in a race from King's Cross to Edinburgh.

This was intended to simulate the situation in
the 1950s, although practical changes to the
roads and the way the railways operate in the
intervening period meant that a true comparison
was impossible. Nevertheless, the programme
was hugely popular and generated further
invaluable publicity for *Tornado*.

Early May saw *Tornado* at the North York
Moors Railway, followed by further main line
charter runs, and then an appearance at the
West Somerset Railway in early June.

In late June there was an emotional return
to the place where the real construction had
started, as *Tornado* starred in a gala weekend of
primarily Great Western locomotives at Tyseley
Locomotive Works, attracting big crowds,
despite being an interloper on GWR metals.

In July and August *Tornado* was back on
the national network, mainly on charter runs
in the South West of England and into South
Wales, including a highly successful series of
weekly runs on the 'Torbay Express', Bristol to
Kingswear and return.

Late August saw the locomotive return to
charter work in the London area, followed by
a maintenance visit to Didcot, during which it
also featured in a special weekend at the Great
Western depot.

September started with a rather emotional
assignment which saw the locomotive at
the head of the 'Winton Train' from Harwich
to Liverpool Street. This was the final leg
of a charter from Prague to London to
commemorate the 70th anniversary of the train

commissioned by the now Sir Nicholas Winton to evacuate 669 Jewish Czechoslovak children just before the outbreak of the Second World War.

Further East Coast Main Line charters followed during September, leading to the next big 'first' for *Tornado*. On Saturday, 3 October, the locomotive headed out of York for its first run over the famous Settle to Carlisle route. Battling against wet and windy weather which later in the day stopped all other traffic on the route, *Tornado* won through, despite dropping time due to the weather conditions. After a quick overnight turn round, she was back on the same route the following day, this time in somewhat better weather conditions.

A week later and *Tornado* was back on the route again on another charter, but this time was routed back over Shap and down the West Coast Main Line for the first time, being taken off the train at Preston.

The following weekend and it was a return to Barrow Hill for the annual covenantors' convention, and then down to the Severn Valley Railway for another hugely successful

ABOVE September 2009, and an emotional moment as No. 60163 *Tornado* approaches Manningtree on its way to Liverpool Street with the 'Winton Train'. *(Don Brundell/A1SLT))*

BELOW Early October 2009, and *Tornado* tackles the famous Settle–Carlisle route – twice in two days! *(Richard Whiteley/A1SLT)*

week on a preserved railway. Further ECML runs followed in November and into December, with the final runs of the year on 21 December, creating yet another headline for *Tornado*. With most of the South East paralysed by snow, and many commuter trains cancelled, *Tornado's* final public run of the year was on a dining train on a Kent circular from London Victoria. With minutes to go before departure, a number of tired and dejected commuters turned up at the ticket barrier thinking this was a service train – they were elated when the operator, Steam Dreams, agreed to take over one hundred of them home. The steam heritage movement undoubtedly gained a few more supporters that night.

The following day, *Tornado* headed back light engine to York for a well-earned rest while winter maintenance was carried out at the National Railway Museum. The first full year of operation, 2009, had been a spectacular 12 months for the locomotive and The A1 Steam Locomotive Trust.

By the time it was resting in the NRM, *Tornado* had clocked up over 17,000 miles since those first moves in steam in July 2008. Of these, some 14,500 miles were in 2009, 12,500 on the main line and 2,000 on preserved railways.

One other major development during 2009 was the introduction of a merchandising operation to produce and sell *Tornado*-branded memorabilia. Following on from early public demand, a small group of covenantors, led by Gillian Lord, and with the full support of the Trust, set about sourcing a wide range of merchandise, to suit every pocket, and to keep this 'refreshed' with new items. This has gone from strength to strength, and the Trust now has a sales stand at all events where *Tornado* is present, merchandise is available on all *Tornado*-hauled excursions, and there is the option to buy on-line through the Trust's website (www.a1steam.com). The operation is made possible by the dedication of a team of volunteers, which enables the significant profits generated from sales to be ploughed back into reducing the residual debt on the build, and to support the on-going running and maintenance of *Tornado*.

BELOW Merchandise director Gillian Lord (in blue) and the team with the sales stand at the Severn Valley Railway in October 2009. *(Neil Whitaker/ A1SLT)*

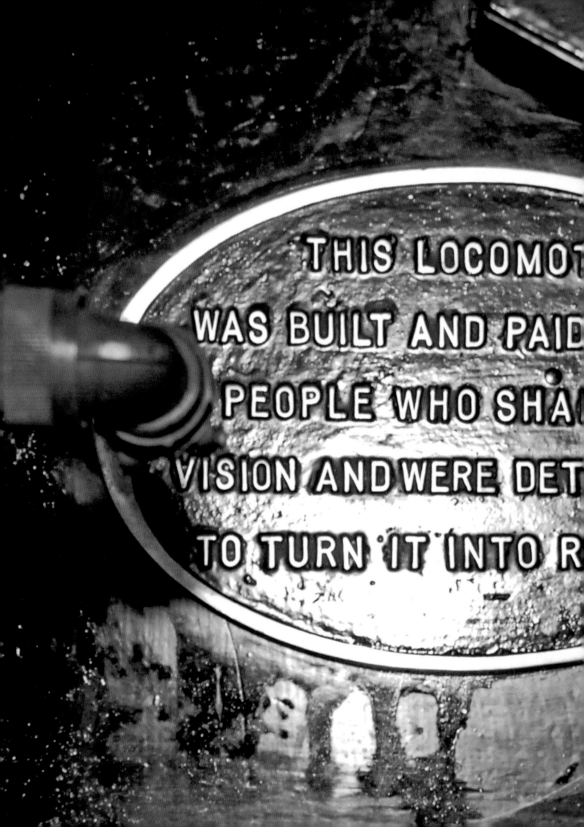

Chapter Eight

And to follow...

Back in 1990, when the first thoughts of a new-build locomotive were starting to crystallise, few if any could have imagined the impact that would be brought about by this project. Many well-known figures in the heritage railway movement were deeply sceptical as to whether the build would ever be completed, and indeed, a number of the early covenantors have admitted that joining the project was a 'bit of a punt'. They were hopeful and committed – but privately doubted they would ever see anything for their investment.

OPPOSITE Mounted on the front frames ahead of the smokebox – the plaque that says it all. *(Geoff Smith/A1SLT)*

Even the most ardent supporters, including those eventually tasked with managing the project to completion, have been taken by surprise by the reception that the locomotive has received wherever it has gone. This enthusiasm is not restricted to the hard core railway enthusiasts, but has spread right through the general public, young and old, male and female, many of whom are too young to remember steam locomotives in regular service, but as a result of this project are now drawn by the lure of steam traction.

During the last 18 months, the heritage steam movement has seen record interest, and a significant contributor to this has been the '*Tornado*' effect'. A number of other 'new-build' projects are now underway, inspired by the success of *Tornado*, including, among others, a North Eastern Railway Class G5 0-4-4T (with assistance being provided by the

Trust's Engineering Director – David Elliott), an LMS 'Patriot' 4-6-0, a BR 'Clan' class 4-6-2, a BR 3MT 2-6-2T, a GWR 'Grange' 4-6-0, and a LNER 'Sandringham' class 4-6-0. With some of these it is planned to use components recovered from scrapped locomotives, although all will involve a substantial new-build element. Some of these builds are intended to create new main line traction, while others are being built for heritage railway use only.

In building *Tornado*, the A1 Steam Locomotive Trust has pulled together a wide range of expertise and skills, which has not been seen in this industry since the demise of regular steam locomotive construction. It has provided training to enthusiastic apprentices, and has welcomed volunteers prepared to give up their spare time just to say they contributed to the building of a new main line locomotive. In addition, it has established professional contacts and relationships with those responsible for running the railway network today. Not least, it has assembled a large group of enthusiastic covenantors (and many are still coming on board), prepared to contribute small amounts each month to ensure that a valuable part of our heritage is retained for future generations.

In order to bring the project to completion in a sensible timeframe the Trust took on significant financial risks in terms of loans, and the first priority on completion was to bring about financial security. To do this, loans had to be repaid, and money set aside for the future running and maintenance costs of the locomotive, part of which comes from the fees earned by *Tornado* in service. This task has been greatly assisted by the continued growth in covenantor numbers, and by the profits from *Tornado*-branded merchandise sales, as well as, of course, the revenue earned by the locomotive itself.

However, keen to capitalise on the experience and expertise which had been gained over the 18 years of hard work, the Trust also set about considering the next step. Early in 2010 it announced that a feasibility study was being undertaken into the possibility of building a Gresley P2 class locomotive. Only six of these, with a 2-8-2 (Mikado) wheel arrangement, were built in the period 1934–36, and by the mid-1940s all had been rebuilt by Thompson as A2/2 class 4-6-2s. They had a mixed success in service, but were considered a big step forward in locomotive design, and had it not been for the death of Gresley in 1941, would probably have been subject to further development and refinement to realise their full potential.

Whether the P2 build will go ahead remains to be seen, but one thing is certain. The huge success of the *Tornado* project has ensured that we will be building steam locomotives in the UK for many years to come.

Appendix 1

Specification – key facts about *Tornado*

Wheel arrangement	4-6-2
Length over buffers (including tender)	72ft 11¾in
Maximum height	13ft 0in
Maximum width (over cab side screens)	9ft 2³⁄₈in
Total locomotive and tender weight (full)	170 tons 16cwt
Weight of locomotive only (full)	104 tons 14cwt
Maximum axle load	22 tons 7cwt
Boiler pressure (maximum)	250lb/sq in
Nominal tractive effort @ 85% boiler pressure	37,397lb
Maximum speed	90mph (currently limited to 75mph by UK railway regulations)
Diameter of driving wheels	6ft 8in
Diameter of bogie wheels	3ft 2in
Diameter of trailing carrying wheels	3ft 8in
Cylinders	x3
Piston stroke	26in
Piston diameter	19in
Valves	Inside admission piston
Valve diameter	10in
Valve gear	Walschaerts (three sets)
Exhaust system	Double-choke Kylchap
Axle bearings	Timken taper roller throughout
Boiler	
Type	Diagram 118A
Maximum diameter	6ft 5in
Overall length	29ft 2in
Distance between tube plates	16ft 11⁵⁄₈in
Heating surface (firebox)	245.30sq ft
Heating surface (small tubes)	1,211.57sq ft
Heating surface (superheater flue tubes)	1,004.50sq ft
Total evaporative heating surfaces	2,461.37sq ft
Superheater heating surface	697.67sq ft
Grand total heating surface	3,141.04sq ft
Firebox	Wide (Wooton) type with combustion chamber
Grate area	50sq ft
Grate type	Rocking with hopper ashpan
Tender No. 783	
Coal capacity	6.5 tons
Water capacity	6,270 gallons
Weight (full)	66 tons 2cwt
Wheel arrangement	8 wheels, rigid frame
Wheel diameter	4ft 2in

Data courtesy *Locomotives of the LNER Part 2A*, published by the RCTS, 1986, modified where appropriate for No. 60163 *Tornado*.

Appendix 2

Timeline of key events

1990 Project conceived and launched.

1991 Start made on collating drawings of the original Peppercorn A1s. Name *Tornado* chosen.

1992 Work on drawings continues, initial enquiries made on obtaining certification; plan anticipates work starting in around 18 months.

1993 Work on drawings continues. Trust continues to attract respected individuals capable of adding significant skills to the project.

1994 Work starts – main frames cut and profiled, cylinder patterns ordered. Trust attracts principal sponsor William Cook plc, which will prove critical to the success of the project.

1995 Work continues on main frames, cylinders cast, patterns made for driving wheels. Nameplates designed and manufactured. Negotiations with Darlington Borough Council result in a deal for the Trust to take a lease on Hopetown Works (home of the old Darlington Carriage Works).

1996 Work continues on the main frames and associated components, including the cylinders. Main driving wheels cast.

1997 Locomotive frames completed at Tyseley. Preparation of Hopetown Works completed and frames moved into the new works.

1998 Smokebox construction begun, including smokebox door. Rest of locomotive wheels cast, and all, including main driving wheels, delivered to Riley's at Bury.

1999 Wheels and bearings fitted to axles, smokebox construction completed. Start made on forging the motion components. Initial cab structure manufactured.

2000 Work continued on the motion components, and further progress made on front bogie. Additional work carried out in Hopetown Works to facilitate later stages of the construction.

2001 Work started to identify a suitable boiler supplier.

2002 Work on motion components continued, work started to reduce cab height to the maximum 13ft 0in required by the Regulations. Orders placed for the cylinder end cover castings.

2003 Cartazzi hornblocks and stays fitted for the rear axle, further work carried out on motion components.

2004 Bond issue launched to fund the boiler manufacture, and boiler supplier selected (Dampflokwerk Meiningen in eastern Germany). Start made on fitting motion components to the locomotive.

2005 Order placed for boiler. Further work carried out on motion components, and platework attached to the main frames. Two sets of brake control and sanding gear obtained, and work started to overhaul these.

2006 Work continued on motion components, including pistons and valves. Pipework well under way. Boiler completed and delivered. Tender construction progressing well.

2007 Motion almost complete, boiler ancillaries fitted, air pumps for braking system delivered and further work carried out on braking system. Wheelsets balanced, and work on tender nearing completion.

2008 A key year! Boiler test successfully completed. Final assembly completed, tender components delivered and assembled. First runs in steam at Darlington (with Press launch), testing completed at the Great Central Railway, followed by successful main line tests. Painting in apple green livery completed, and unveiling ceremony to covenantors.

2009 *Tornado* enters service to universal acclaim. Formal naming ceremony at York station by Their Royal Highnesses, the Prince of Wales and the Duchess of Cornwall. The locomotive pulls unparalleled crowds wherever it goes – main line and heritage railways. The 'Tornado effect' takes a firm hold.

Appendix 3

Official sponsors

Whilst a large part of the financing of *Tornado* has come from the army of covenantors who have patiently donated small amounts of money each month, some over the full lifetime of the project, successful completion would not have been possible without the support of a smaller number of commercial sponsors.

These organisations have provided a combination of cash, expertise, and the supply of goods and services at very favourable rates, with the aim of seeing the recreation of a small part of our proud industrial heritage. The Trust is eternally grateful to these organisations for their support.

Principal sponsor:
William Cook Cast Products

Other sponsors:

AB Hoses & Fittings Limited
British Aerospace (now BAe Systems)
British Steel Distribution
British Steel (now Tata Steel)
Craftmaster Paints
Darlington Borough Council
East Midlands Trains
Footplate Equipment Limited
GNER
Gore Tex
Handpoint Limited
Head of Steam, Darlington Railway
  Museum
Hima-Sella
I. D. Howitt Limited
M. H. Spencer Limited

Marshall of Cambridge Aerospace
McCready Steel Stockholders
National Express East Coast
New Cavendish Books
Odd Bolt
ONE North East
RAF
Redcliffe Imaging Limited
Rock Oil
Rolls-Royce
Steam Railway magazine
Tasque Consultancy
Timken
Total
Unipart Rail
Virgin Trains

OPPOSITE Artists impression of Tornado in the liveries of the original Peppercorn A1s, and which Tornado will carry over the first ten years in service – apple green, BR blue and Brunswick green (with both the BR emblem and the later BR crest on the tender sides). *(Copyright in these images rests with The A1 Steam Locomotive Trust)*

Index